A Thread
Across the
Ocean

A Thread Across the Ocean

Ocean

The Heroic Story
OF THE
Transatlantic Cable

John Steele Gordon

SIMON & SCHUSTER

First published in Great Britain by Simon & Schuster UK Ltd, 2002
A Viacom Company

1 3 5 7 9 10 8 6 4 2

Simon & Schuster UK Ltd
Africa House
64–78 Kingsway
London WC2B 6AH

www.simonsays.co.uk

Simon & Schuster Australia Sydney

A CIP catalogue record for this book is available from the British Library
Hardback ISBN 0–7432–0809–9
Trade paperback ISBN 0–7432–3127–9

Printed and bound in Great Britain by
The Bath Press, Bath

Every effort has been made to locate and contact all the holders of copyright to material
reproduced in this book.

Illustrations on pages iii and 207 are used by permission of The Granger Collection,
New York. Illustrations on pages 11 and 34, courtesy of the New York Historical Society,
New York City. Illustration on page 37 is used by permission of the Library of Congress.
Illustration on page 43, courtesy of the Chamber of Commerce, New York City.
Illustrations on pages 81, 103, and 105 are from Harper's Weekly [Telegraph
Supplement] September 4, 1858. Illustration on page 124 is from Harper's Weekly,
August 14, 1858. Illustrations on pages 133, 136, and 204 are used by permission of
Culver Pictures. Illustrations on page 137, courtesy of the Burndy Library, Norwalk,
Connecticut. Illustrations on page 148, courtesy of the Nautical Photo Agency.
Illustration on page 156 is by Ley Kenyon published in Builder and Dreamer by Laurence
Maynell, The Bodley Head Ltd.,k London 1952. Illustration on page 161, courtesy of C.
C. Bennett. Illustration on page 175 used by permission of The Illustrated London News
Picture Library. Illustration on page 194 is used by permission of The Metropolitan
Museum of Art, all rights reserved. Illustration on page 199 is courtesy of the National
Archives of Canada/C–004484.

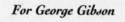

For George Gibson

In comparison with the ship from which it was payed
out, and the gigantic waves among which it was
delivered, [the cable] was but a mere thread.

—NICHOLAS WOODS
The Times of London , 1858

Let us then be up and doing,
    With a heart for any fate;
Still achieving, still pursuing,
    Learn to labor and to wait.

—HENRY WADSWORTH LONGFELLOW
"A Psalm of Life"

CONTENTS

A NOTE ON MONEY

*T*he laying of the Atlantic cable was one of the great international undertakings of the mid-nineteenth century. Although organized and pushed to completion by an American, much of the technology and most of the financing came from Britain. Thus, figures in both dollars and pounds appear frequently on these pages. What was the relative value of each currency at that time?

In the middle of the nineteenth century, Great Britain had the world's largest economy, the pound sterling was the dominant currency of international trade, and the Bank of England was, effectively, the world's central bank. Since the early eighteenth century, the pound had been valued at three pounds, seventeen shillings, ten and a half pence to the ounce of gold. After 1821 the Bank of England had stood ready to buy or sell unlimited amounts of sterling for gold at that rate. This was the beginning of the "gold standard" that would spread around the world over the next half century and endure in one form or another until 1971. As a result, Bank of England pound notes traded at par with gold and were acceptable as payment almost everywhere.

The dollar had been valued at $20.66 to the ounce of gold since 1834, but the United States had no central bank after 1836, when the charter of the Second Bank of the United States expired and President Andrew Jackson vetoed its renewal. The United States Treasury minted gold and silver coins, but American paper money, before the Civil War, was issued by thousands of state-chartered banks. These notes traded for gold at ratios that ranged from nearly par to nearly infinity, depending

on the reputation of the bank and the distance of the transaction
from the bank's headquarters, where they could be redeemed for
specie (in coin). American paper money had no currency abroad.

With the advent of the Civil War, the United States
Treasury began issuing "greenbacks," paper currency that was not
redeemable in gold (and not acceptable in payment of federal
taxes). The ratio of the value of these greenback dollars to the
gold dollar tended to correlate, inversely, with the fortunes of the
Union Army, reaching a high of 2.87 greenback dollars needed to
buy 1 gold dollar just before the Battle of Gettysburg in 1863.

That year the National Banking Act ended the issuance of
bank notes by state banks and provided for the chartering of
national banks, which could issue banknotes only under strict reg-
ulations and to a uniform design. This gave the United States a
sound and coherent paper money system for the first time in its
history. With the return of peace and of the Confederate states
to the Union, the dollar became once more fully convertible into
gold in 1879.

Based on their values relative to gold, in the 1850s and 1860s,
one pound was worth slightly more than five dollars. The various
monetary sums given in this book can be accurately converted at
this ratio.

ACKNOWLEDGMENTS

I wish to thank my publisher, George Gibson, who had the idea that I should write this book and waited more patiently than I would have for it to be delivered; Katinka Matson, my agent; Jackie Johnson, who edited the book; my friend David Stebbings, who made sure I got the technical details correct; Judith Gelernter of the Union Club Library, for allowing me to keep books long after they were past due; Timothy Field Beard for details on the ancestry of the Field family; and John K. Howat for information on the South American adventure of Cyrus Field and Frederic Church.

A Thread Across the Ocean

I

"AN ENTERPRISE WORTHY OF THIS DAY OF GREAT THINGS"

THOMAS NIGHTINGALE HAD PROSPERED in South Carolina almost from the day he had arrived as a young man from his native Yorkshire in the 1720s. He got his start operating a cow pen on the frontier but soon branched into numerous other activities, including building Newmarket Race Track in Charleston and importing some of the first thoroughbred horses to the North American colonies.

The timing of his arrival had been fortunate. Trade in rice and indigo was fast making the Carolina tidewater the richest part of the British North American empire, and Thomas Nightingale grew rich right along with his adopted land.

In 1760, already well established among Charleston's aristocracy, he decided to add one more proof of his status in that very status-conscious society. He bought a pew, number 101, in St. Michael's Church, then under construction in Charleston. With the great wealth at the congregation's disposal, little expense had been spared in the building of St. Michael's, a masterpiece of American colonial architecture. Much of the woodwork, for instance, would be supplied by Thomas Elfe, the city's leading cabinetmaker and himself a pew holder. A pew in such a church did not come cheaply. But for fifty pounds—more than a workman's annual wage in the middle of the eighteenth century—Thomas

Nightingale received a deed to the pew, signed and sealed by the church commissioners.

That deed, to our eyes, has one very curious aspect. It is dated "the fifth day of December in the Year of our Lord One Thousand Seven Hundred and Sixty and in the Thirty-Fourth year of the Reign of our Sovereign Lord King GEORGE the Second . . ."[1] But George II had died suddenly of a burst blood vessel on October 25, 1760, while in his water closet. December 5, therefore, was in fact in the first year of the reign of his grandson, King George III.

It is a measure of the perceived vastness of the Atlantic Ocean in the eighteenth century that the king's richest North American possession remained ignorant of his death a full six weeks after the event. Charleston, in fact, did not learn of the king's death for another two weeks or more.

Yet America's civilization and character developed during the colonial period in the context of this profound isolation from its European roots. In 1620, the *Mayflower* had sailed from Plymouth, in Devonshire, on September 16, and raised Cape Cod only on November 9. That was considered a very good passage at the time, and in fact it was still a good passage two hundred years later. It was by no means unprecedented for a ship unfortunate in weather to take four months to make the trip from the Old World to the New.

Because the trip was so long, expensive, and perilous, only a handful of immigrants to the New World—mostly members of the colonies' business and political elites—ever had the opportunity to return to the Old. Thus, to set sail for America in the seventeenth and eighteenth centuries was, almost always, to say good-bye forever to all the emigrant had known and loved.

Since for all practical purposes news could travel no faster than human beings could carry it, knowledge of events in Europe—the center of the Western world—was just as slow to flow across the ocean as men and goods. North America was not only three thousand miles from Europe—it was two months from it as well. There was not even a regular postal system; letters were entrusted to anyone willing to carry them, to be delivered when and if possible.

Today, such isolation is almost inconceivable. After all, it took the Apollo astronauts only three days to reach the moon, a distance almost a hundred times as great as the width of the Atlantic, and news from the moon (not, to be sure, that there is much) could reach us in seconds.

But in Thomas Nightingale's day, the vast gulf between Europe and America was simply a fact of life. Like growing old, or needing to sleep for several hours a day, it was taken as a given, if sometimes regretted.

But even as King George lay dying in his water closet, many of his subjects, especially in the Midlands of England, were already deeply involved in a process that would profoundly alter the boundaries of what was possible. The cloth industry, for example, a mainstay of the British economy for centuries, had begun to mechanize, adopting the factory system of manufacture, which would come to dominate the world economy. John Kay's flying shuttle, introduced in 1733, considerably increased the speed with which cloth could be woven. The spinning jenny and the water frame, introduced in the 1760s, greatly accelerated the manufacture of yarn. The power loom in the 1780s completed the mechanization of the industry.

At first these machines were powered by falling water pushing on mill wheels. Then, in 1769, the Scotsman James Watt patented a greatly improved steam engine, and in 1784 introduced a rotary version, capable of turning a shaft. The new power source, which made work-doing energy both cheap and capable of being applied in almost unlimited amounts to a single task, proved the catalyst of profound change. Dubbed the Industrial Revolution in 1848 (when it was already almost a century old), it swept away the world of Thomas Nightingale and George II in a matter of two generations and created the modern world.

Politics helped. With the final defeat of Napoleon at the Battle of Waterloo, on June 16, 1815, the Western world entered into a period of peace that would last nearly a hundred years, until the outbreak of the First World War on August 1, 1914. In this period of relative peace, with the exception of the American Civil War, wars were mostly short and often distant from the centers of Western civilization. Thus, national and individual energies could be directed to peaceful pursuits. With steam, the new energy source, offering a myriad of possibilities, men of the highest talent and ambition — men who in earlier times might have gone into politics, the military, or trade — moved to exploit these possibilities to the fullest.

Soon the steam engine was being used not only to power factories that produced an ever-growing number of products at prices the middle class could afford but also to revolutionize transportation. In the eighteenth century, bulky goods were transported by water or they did not move at all. But in the first decade of the nineteenth century, Richard Trevithick in England and Oliver Evans in the United States developed high-pressure steam engines that produced far more power per unit of weight than Watt's engine had. Fitted to wagons that moved on rails, a technique used in mining since the sixteenth century, the new high-pressure engines were capable of hauling large loads at unprecedented speeds over land.

After George Stephenson solved any number of engineering puzzles and built the Manchester and Liverpool Railroad, which began operating profitably in 1829, railroads quickly spread like a spiderweb across both Europe and North America. This development greatly reduced the cost of transporting both people and goods, and made possible markets of national scope. This, in turn, greatly reduced the cost of manufacturing those goods through economies of scale—decreasing their price and increasing the demand for them. The rate of economic growth and wealth creation soared.

This new wealth enormously increased the size and influence of the middle class. Indeed, the very term *middle class* was coined only in 1812. And this new socioeconomic group began to enjoy a lifestyle quite unknown even to the rich fifty years earlier. In the first half of the nineteenth century, most of the attributes of the modern domestic world became common in their households: central heating, running water, and abundant interior lighting; cheap newspapers, magazines, artwork, and books; cheaper clothes, linens, china, carpeting, and wallpaper.

Given human nature, it is not surprising that people indulged in these rich new possibilities, and an overstuffed, densely furnished, intricately appointed style of decor was the height of fashion in the mid–nineteenth century. At that time, more was more.

This sudden, accelerating increase in the standard of living and in the technology of everyday life also affirmed a belief in progress and in the possibility of improvement, a sense that anything—economic, technical, or social—was possible. The nineteenth century thus became a great age of both optimism and reform as people began to tackle social problems that had once been thought intractable artifacts of the human condition, such as poverty, drunkenness, and gambling. Not the least of these reforms was in politics.

Where wealth, and thus power, had once been concentrated among the owners of land and the great merchants, now a new class of factory owners rose who were often richer—and more liquid—by far than the older moneyed class. In the 1830s the franchise was widened in both Britain and the United States, and in 1832 Parliament was reformed and the seats redistributed for the first time since the time of Henry VIII. This recognition of the new economic reality via a peaceful, constitutional transfer of power from the landed aristocracy to the bourgeoisie is one of the most remarkable events of the nineteenth century or, indeed, of any age.

The wealth these factory owners piled up changed society in other ways. Money became an obsession. *"Wealth! Wealth! Wealth!"* wrote the English social critic John Sterling as early as 1828. *"Praise to the God of the nineteenth century! The Golden Idol! The mighty Mammon!* Such are the accents of the time, such the cry of the nation. . . . There may be here and there an individual who does not spend his heart in laboring for riches; but there is nothing approaching to a class of persons actuated by any other desire."[2] The year before, the young novelist and future prime minister Benjamin Disraeli had coined the word *millionaire* to describe the members of this new class.

The steam engine was also soon applied to transportation at sea. The *Savannah*, built in New York in 1819, weighed 320 tons and was equipped with a 90-horsepower engine and collapsible paddle wheels. That year she crossed from Savannah, Georgia, to Liverpool in only 27 days and 15 hours (663 hours), having used her engine for about 80 hours of the trip.

Measured in time, the Atlantic was shrinking rapidly before the onslaught of the Industrial Revolution, but transportation was by no means the only field in which speed was increasing by orders of magnitude, thanks to another profoundly new technology. Long-distance communication as well was changing beyond recognition.

———

IT HAD NEVER BEEN QUITE TRUE that news and men moved at the same speed. Smoke and flag signals had been around since antiquity. The homing instinct of pigeons was exploited to carry messages at least as early as A.D. 1150, when the sultan of Baghdad established a pigeon postal system. In 1588, Queen Elizabeth I ordered the construction of bonfires along the south coast of England, to be ready to convey to London in a couple of hours instead of days a single message: "The Armada has been sighted."

The French took visual signaling to the ultimate. Beginning in 1794, Claude Chappe built a series of semaphore stations between Paris and important military posts, such as the great naval base at Brest, three hundred miles west of Paris on the tip of the Brittany peninsula. Equipped with a mast and two cross arms, these stations were spaced five to ten miles apart and worked by four or five people each. They were, in effect, a series of gigantic Boy Scouts wigwagging to each other across the French landscape. By this method, messages could be transmitted at a rate of several hundred miles a day, far faster than any other method of the time.

The system had numerous disadvantages. For one thing, it was so expensive on a per word basis that only a sovereign government could afford to utilize it. For another, it was useless in bad weather and at night. And because each message had to be copied and repeated many times by fallible human beings, the error rate was extremely high. Regardless, timely information can be so valuable that the Chappe system (which he dubbed a *telegraph*—from the

Greek words meaning writing at a distance) and variations of it spread rapidly across Europe.

Even before Chappe developed his signaling system, however, indeed before Thomas Nightingale bought his church pew, the basis of a wholly new communications technology was being laid by—to use the marvelous phrase of Sir Arthur Clarke—"beachcombers on the further shores of theoretical physics."[3]

That basis was electricity. It had first been recognized as a separate phenomenon in the seventeenth century, although its nature was an utter mystery at the time. It was in the eighteenth century that that nature began to be studied seriously. In 1747 Sir William Watson demonstrated that it was possible to transmit electricity down a metal wire for long distances. In the next decade Benjamin Franklin explored the subject. He proved among much else that lightning was an electrical phenomenon, thanks to his famous— and *extremely* hazardous—experiment with a kite and a key. It was also Franklin who coined many words that are still used in discussing electricity, such as *positive, negative, battery,* and *conductor.* He once amused dinner guests by igniting spirits of alcohol from across the Schuylkill River in Philadelphia, using an electric wire to convey the spark.

In 1753 an unknown writer, using only the initials C.M., proposed in *Scots' Magazine* the use of electricity to convey information over long distances by means of wires. The first such system actually built, in Geneva in 1774, used one wire for each letter of the alphabet. The current would charge a pith ball with static electricity, which in turn would attract a bell and ring it. This alphabetical carillon actually worked, after a fashion, but was hardly a practical system. It was only with the invention of the electromagnet and much better batteries, and the availability of cheap wire thanks to the Industrial Revolution in the early years of the nineteenth century, that it became possible to convert the electric telegraph from a parlor trick to use as a world-transforming technology.

Like many seminal inventions of the industrial era—the railroad, automobile, television, the computer—the electric telegraph was not the invention of one person. Several people contributed vital parts of the system, but it was William Fothergill Cooke and Charles Wheatstone in England and Samuel F. B. Morse in the United States who first built practical, working telegraph lines of substantial length and transmitted messages over them. Eventually it was Morse's particular system and, especially, his marvelously efficient code—the only part of his telegraph system wholly original with Morse—that became the standard.

The telegraph was something very new under the sun, something that would have been utterly inconceivable to the world of Thomas Nightingale. The telegraph could transmit information at very high speed—thousands of times faster than it could be physically carried and hundreds of times faster than Chappe's visual telegraph could transmit it—and at very low cost. So it is not surprising that once its practicality was demonstrated, the telegraph spread with astonishing speed, often using the convenient pathways forged by the equally fast-spreading railroads.

The relatively small countries of western Europe were mostly wired by 1850. Even the vast and only partially settled United States had a telegraph line running across the continent to California by 1861, only seventeen years after Morse had first tapped out the dashes and dots spelling "What hath God wrought?"

Because nearly three-quarters of the surface of the Earth is covered by water, it was not long before men began to consider the problem of extending the miracle of telegraphy across these bodies of water. Indeed, Samuel Morse studied the problem in 1842, two years before he sent his famous message from Washington to Baltimore, when he laid an insulated copper wire across New

York harbor and transmitted an electric current through it. In 1845, Ezra Cornell (who would found Cornell University two decades later) laid a line that crossed the Hudson River from Fort Lee, New Jersey, to Manhattan, where the George Washington Bridge is today. It worked satisfactorily for several months until ice broke it.

Stringing wires on land and across rivers was one thing, however, and crossing a substantial body of water quite another. But because the richest, most powerful, and most technologically advanced country in the world in the mid–nineteenth century happened to be located on an archipelago off northwest Europe, it was not long before the attempt was made. In 1845 two English brothers, John and Jacob Brett, sought government approval for laying a telegraph line across the twenty-two miles that separated the island of Great Britain from the coast of France. Finally receiving permission from the two governments in 1850, they ordered twenty-five miles of insulated wire no bigger around than a man's little finger, set the seven- by fifteen-foot spool on the back of a small steam tug, and set off across the channel, reeling out the wire as they went.

The cable was so light that it wouldn't sink, and thirty-pound lead weights had to be added every so often. The Bretts reached France by nightfall and quickly tried to send a message back to England. What emerged at Dover was gibberish, but at least they had proved it was possible to transmit electricity across a submarine cable of that length. As *The Times* of London expressed it, "The jest of yesterday has become the fact of today."[4]

The next morning, however, the line was completely dead. It was subsequently discovered that it had been fouled by a fisherman's anchor. It was so light he was easily able to haul it up and, thinking the copper core might be gold, he cut out a piece to show others.

Regardless of the failure of the world's first submarine telegraph cable, the next year a new and vastly improved one was laid, and it worked. England was connected for the first time since the last Ice Age to the continent of Europe, so news in Paris could be news in London at the same time. Soon submarine cables connected England to Holland and Ireland and, in 1854, one was laid from Italy to Sardinia and Corsica. But these were baby steps, across shallow and narrow seas.

Then, in the same year the cable between Corsica and Sardinia was laid across the nine-mile-wide Strait of Bonifacio, an American businessman named Cyrus Field made the astonishing

Cyrus Field

decision to lay one across the Atlantic Ocean and end at a stroke his country's remoteness from the rest of the world.

At that time, the longest submarine cable in operation ran 110 miles and lay at a depth no greater than 300 fathoms. An Atlantic cable would have to be more than 2,000 miles long and reach depths of 2,600 fathoms. It was a bit as if someone in the 1950s, reading of the success of the Russian *Sputnik*, had decided to organize a manned expedition to Mars.

That Field would get to his Mars only after twelve years, five attempts, and endless trials and tribulations is hardly surprising. But get there he did, and the attempt was entirely in keeping with both the man and his times. In the words of Salmon P. Chase, secretary of the treasury and chief justice of the United States, it was certainly "an enterprise worthy of this day of great things."[5]

His epic quest would fascinate the world he lived in and deeply involve some of the finest scientific, engineering, and business minds of that, or for that matter any other, day. But it was Cyrus Field alone who made it happen, for he served the same function in the enterprise of the Atlantic cable that a producer serves in a theatrical production. A producer does not act or direct or design scenery. But without him, neither does anyone else.

By the mid-nineteenth century it was known that the powers of science, technology, and wealth—all greatly increased by the Industrial Revolution—were capable of profound synergy. And some people were beginning to realize that synergy could be maximized, even multiplied, if the various elements were brought together and utilized in a deliberate and productive way by a new type of businessman, the entrepreneur. Indeed, the very word *entrepreneur* entered the English language only in 1852, and Cyrus Field is a nearly perfect exemplar of the new breed. The concept was, perhaps, the greatest of the myriad inventions of the age in which Cyrus Field lived and the greatest gift that age bestowed upon ours, for it was nothing less than the key to the future.

Only much money, faith, dogged perseverance, innovation, and occasional great good luck would see him and those he brought together through to one of the true triumphs of his era.

II

CYRUS FIELD

IN THE EARLY DAYS of the American nation, several New England families each produced an extraordinary number of people of high achievement and lasting fame. Over the course of four generations, the Adams dynasty produced two presidents and a number of diplomats and writers whose names are still familiar. Harriet Beecher Stowe, who wrote the most politically influential novel in American history, *Uncle Tom's Cabin,* was the sister of Henry Ward Beecher, the best-known (and probably the best-paid) American minister of the nineteenth century. His Plymouth Church in Brooklyn, New York, held twenty-five hundred people and was standing room only whenever he preached.

Cyrus Field was born into another such family, perhaps the most prolific of all. Zechariah Field, grandson of the astronomer John Field—who introduced the heliocentric theory of Copernicus to England—arrived in the New World as early as 1629. Several generations later, his descendant David Dudley Field, Cyrus's father, was born in 1781 in East Guilford (now Madison), Connecticut.

David Field was born to be a minister. Even as a boy he would preach to his friends (at least for as long as they would listen). After graduating from Yale with high honors in 1801, he studied under Dr. Charles Backus, of Somers, Connecticut, whose book *The Nature and Influence of Godly Fear* had been widely read. He was soon licensed to preach himself and the next month he married Submit Dickinson, known as "the Somers beauty."[1]

The Reverend Field was a New England preacher of the old school, "tall, erect of carriage with keen piercing steel-gray eyes set in his typically Puritan, stern, Old-Testament type of face."[2] As one might expect, he was not given to mincing words. When Peter Lung was sentenced to hang for having hacked his wife to death in a drunken rage, Field preached a sermon at his execution, as the condemned stood before him. What you have done, he told Lung, "subjects you to ignominious execution. From this there is no escape. . . . Pray God, then, if perhaps your sins may be forgiven you. Cry to him, God be merciful to me, a sinner! And continue those cries till death shall move you hence."[3] But at least the Reverend Field held out a modicum of hope. "Before yonder sun shall set in the west," he told the condemned man, "your probationary state shall be closed forever. This day you will either lift up your eyes in hell, being in torment, or, through the rich, overflowing, and sovereign grace of God, be carried by the angels to Abraham's bosom."[4]

The Reverend Field believed fully in the notion that idle hands are the devil's tools. Certainly his own hands were anything but idle. He was a full-time clergyman, first in Haddam, Connecticut, where the seven oldest of his children were born, and then Stockbridge, Massachusetts, where the last three arrived, including Cyrus, the seventh of his eight sons. At that time, the local Congregational minister was among the most important people in every New England community, constantly consulted and involved on matters ranging far beyond religion. But Field also found time, besides writing a weekly sermon—many of which were published—to write numerous books on local history that remain useful source material today. He also served as his Yale class historian and wrote and published his memoirs.

Despite these demands on his time, the Reverend Field took care to school his children carefully in the Bible, the Greek and Latin authors they would not encounter in the local school— which offered little more than the basics of reading, writing, and arithmetic—and English literature. His sons long remembered a favorite line of their father from John Bunyan's *The Pilgrim's Progress,* "To *know* is a thing which pleaseth talkers and boasters; but to *do* is that which pleaseth God."5 There is no doubt that the Reverend Field instilled that idea in his sons.

But while David Dudley Field was a stern preacher and hard taskmaster, he was also a loving father and a good friend who knew how to enjoy himself. Ministers would come to the house for theological discussions over wine and brandy. His son Henry remembered how after the discussions, "peals of laughter . . . shook the place" when his father told stories, and the Reverend Field "laughed till the tears ran down his cheeks."6

Field's professional career was distinguished enough to earn him an entry in the *Dictionary of American Biography,* a standard work of twenty-four volumes, published in the 1920s and 1930s, modeled on the British *Dictionary of National Biography,* one of the great monuments of Victorian scholarship. But the domestic accomplishments of his wife and himself are quite as impressive as his professional ones. No fewer than four of his eight sons (two of whom did not live long enough to make a mark on the world) would also find places of their own in the *Dictionary of American Biography.* That, perhaps, is a record.

His namesake, David Dudley Field Jr., born in 1805, became one of the most highly paid lawyers in the country, representing clients as varied as "Boss" Tweed of Tammany Hall, the Erie Railway, and Samuel J. Tilden, Democratic candidate for president in 1876. But his greatest achievement was the reform of the common law. The Field Code of Civil Procedure, largely his work, was adopted in New York State in the 1840s to simplify the procedures — and eliminate the abuses — of the law that were depicted so savagely in Dickens's *Bleak House*, published in 1852.

The Field Code was quickly adopted in other states and by the federal government. In 1873 it was used as the basis of reforming the law in England and then throughout the British Empire. No nineteenth-century lawyer or judge of the English-speaking world had a greater influence on the law as it is still encountered daily by ordinary people than David Dudley Field Jr.

Stephen Johnson Field (born in 1816) also entered the law. In 1849, however, he joined the great migration to California after gold was discovered there. But he never prospected. Instead he served as a lawyer, invested shrewdly in land, and rose quickly in the state's politics. In 1863 President Lincoln appointed him to the United States Supreme Court, a seat he would hold for a then record thirty-four years.*

Henry Martyn Field, the youngest of the family, born in 1822, became a minister and author like his father. His books on travel, history, and biography proved very popular, often running through several editions and remaining in print for years. His *History of the Atlantic Telegraph*, perhaps his most popular book, was in print for over forty years and is, as we shall see, a prime, indeed inescapable, source for the history of the venture.†

The two remaining sons (not given space in the *DAB*) were

---

*His sister's son, David J. Brewer, also sat on the high court, for twenty years, the first seven of them concurrently with his uncle.

†Henry would marry Henriette Desportes, a Frenchwoman acquitted of murdering a duchess after a spectacular trial in Paris in the 1840s. In 1938, the story was told as a best-selling novel by Rachel Field, a granddaughter of Matthew Field, called *All This and Heaven Too*, which was made into a highly successful movie of the same name starring Bette Davis and Charles Boyer.

only slightly less successful than their brothers. Jonathan Field served three terms as president of the Massachusetts State Senate. Matthew Field became a civil engineer responsible for railroads and bridges in many areas of the country. His railroad bridge across the Cumberland River at Nashville, Tennessee, was destroyed in the battle for that city in the Civil War.

And then there was Cyrus West Field. Born November 30, 1819, Cyrus was less studious than some of his older brothers, who had attended nearby Williams College, but was no less active, ambitious, or intelligent. At the age of sixteen he persuaded his father to let him seek his fortune in business. He made his way from Stockbridge to the Hudson River and took a boat south to New York City. It was a fateful encounter. Cyrus Field would turn out to be a very gifted entrepreneur, and nineteenth-century New York, where he would live for most of the rest of his life, was devoted to moneymaking more than any other city in the nation.

Founded by the Dutch two centuries earlier to facilitate their trade in beaver skins, New York had always been commercially minded. Indeed, the Dutch West India Company was so bent upon profits that it didn't even get around to building a proper church on Manhattan for seventeen years. The Dutch Reform minister simply preached in the fort. Before the end of the Dutch era, which lasted only forty years, New York merchants were trading with Europe and the West Indies and as far afield as the Indian Ocean.

The city became even more devoted to commerce with the lifting of British restrictions on American economic activity at the end of the Revolutionary War. In 1784 the city's first bank opened and New York merchants dispatched the first American voyage to the Far East and began the American China trade. In 1819 New Yorkers began the first regularly scheduled passenger service to Europe and back, which helped mightily to cement the city's position as the country's leading port.

The Industrial Revolution provided the city's businessmen with even greater opportunities, and they seized them. Eli Whitney's cotton gin made cotton king in the South. But the cotton trade with Europe—the fiber would be the country's leading export until the 1930s—was largely brokered through New York.

New York had surpassed Philadelphia as the country's largest city by 1790 and increased its lead in every census thereafter. But though New York was the country's leading city from the early days of the Republic, it became the country's dominating city only after the opening of the Erie Canal in 1825. Once Governor DeWitt Clinton pushed it through an often reluctant legislature, the canal was built in only eight years and proved the most consequential public work in the nation's history.

Before the canal, the vast, burgeoning trans-Appalachian area of the country could not profitably export the rapidly increasing produce of its rich farmland over the mountains to the eastern seaboard. Thus the West had no choice but to utilize the Mississippi to export through New Orleans, or the Great Lakes and the St. Lawrence River, portaging around Niagara Falls.

The Erie Canal provided a direct link between Lake Erie and the Hudson River at Albany, 150 miles north of New York. The canal captured much of the western commerce immediately. Within a few years the Boston physician and poet Oliver Wendell Holmes (father of the Supreme Court justice) would complain that New York had become "that tongue that is licking up the cream of commerce and finance of a continent."7 He was not overstating the case. In 1800, about 9 percent of the country's foreign commerce passed through New York. By 1860 the figure was no less than 62 percent.

Canals became an investment craze, and they were soon joined in popularity by the new railroads. Wall Street enjoyed its first great bull market in the 1830s, and "Wall Street" came to be synonymous with the American financial world.

Thus, thanks to the Erie Canal, the city that Cyrus Field entered for the first time in 1835 was in the midst of an unprecedented building boom. Indeed, New York at that time was the greatest boomtown the world has ever known, rushing pell-mell up Manhattan Island and adding on average ten miles of developed street front per year. While thousands of buildings a year were being built, the streets were clogged with construction materials, and traffic was already notorious for obstruction and delay.

In this remarkable hustle and bustle, the opportunities for making a fortune were nearly unlimited. In 1828, when the city's population was 185,000, there were only 59 New Yorkers thought to have a net worth in excess of $100,000 — a large fortune by the standards of the day. Seventeen years later, the city's population had doubled to 371,000, and the number of citizens worth $100,000 had quintupled, to 294.

In 1835 New York was still largely a city of brick and wood town houses. The brownstone that would come to dominate the domestic cityscape in the next decades was only beginning to be utilized in large quantities. But while New York as a whole was impressive, there were few buildings of note besides City Hall. For example, the present Trinity Church, by now an enduring symbol of Wall Street, would not replace its modest predecessor until the 1840s. Nor was New York much loved by its inhabitants. It was simply changing too fast for that. "Why should it be loved as a city?" *Harper's Monthly* asked in 1856. "It is never the same city for a dozen years altogether. A man born in New York forty years ago finds nothing, absolutely nothing, of the New York he knew. If he chances to stumble upon a few old houses not yet leveled, he is fortunate. But the landmarks, the objects which marked the city to him, as a city, are gone."[8]

Contributing to the disappearance of old New York was the great fire of 1835 on the night of December 16. The night was so bitterly cold that firemen had to chip through the ice on the East River to obtain water and they could do little to halt the spread of the wind-whipped flames. The glow was seen in the night sky of Poughkeepsie, New Haven, and Philadelphia. A total of 674 buildings were destroyed in the area bounded by Wall, Broad, and South Streets, the heart of the old business district.

The sixteen-year-old Cyrus was greatly excited by the fire, writing his parents proudly that he had been up all night helping to fight the conflagration and had spent the previous Sunday "on duty with [his brother] David as a guard to prevent people from going to the ruins to steal property that was saved from the fire and laying in heaps on the streets."9

Cyrus's brother David was already well established as a lawyer in New York, and he arranged for Cyrus to apprentice at A. T. Stewart's store at 257 Broadway, across the thoroughfare from City Hall. Stewart's was the city's leading dry goods store and Stewart would in later years go on to build first the Marble Palace at 280 Broadway, probably the largest store in the United States at the time, and then the Iron Palace at Broadway and Tenth Streets, occupying an entire city block. A genius in the art of retailing, Stewart more or less invented the department store and became one of the city's richest men, leaving an estate at his death in 1876 of more than $40 million.*

Cyrus was paid fifty dollars for his first year's apprenticeship, and his salary was doubled for the second year and doubled again for the third year. He impressed his fellow clerks and his bosses with his friendly, outgoing personality and his attention to business. He carefully noted Stewart's merchandising techniques and never hesitated to ask questions when he wanted to know the reason for something. He also studied on his own, taking night courses in bookkeeping. And, though not involved personally, Cyrus had a ringside seat at the financial crash of 1837, when as many as seven out of ten businesses failed, at least temporarily, in New York City. Strong, well-capitalized businesses survived; badly managed ones did not. It was a lesson he would not forget.

---

*In one of the more bizarre crimes in New York's history, Stewart's body was snatched from its tomb two years after his death and held for ransom. Redeemed for $200,000, it was returned to the Episcopal Cathedral in Garden City, Long Island, a suburb that Stewart had developed.

Stewart, like Field's father, was a stern but fair taskmaster. He expected his clerks to be on time and fined those who were late. The fines accumulated in a fund that Stewart said could be donated to the charity of the clerks' choice. Field was elected treasurer of this fund.

When the time came to select a charity, Field told the clerks to meet him at a local oyster bar so that they could decide on what to do with the money. The first decision they made was to pay the oyster bar's bill out of the fund. Field then suggested that the most deserving group in New York that he could think of to receive the balance of the fund were the clerks at A. T. Stewart's store. His fellow clerks happily agreed and proceeded to divide the money among themselves.

Stewart, of course, found out about this and called a meeting, where he asked Field to give an account. Field told the truth and Stewart only said, pointedly, that in future the firm would be "very careful"[10] in selecting the charity. Apparently he was more amused than angered by the clerks' action, as indicated by the fact that he raised Field's salary to $300 instead of the scheduled $200 in the last year of his apprenticeship.

When Field's apprenticeship came to an end, he decided to strike out on his own, although Stewart made him an attractive offer. His brother David assured their parents that Cyrus was leaving Stewart's "with the best of testimonials of esteem from all his employers and associates."[11] Indeed, his fellow clerks gave him a dinner at Delmonico's, New York's leading restaurant, and one of Stewart's partners presented him with a diamond stick pin.

Many paper mills had opened in recent years along the Housatonic River, which flowed through Field's native Stockbridge, Massachusetts. His older brother Matthew had become a partner in one of these mills but needed help with the business end of things. He agreed to pay his younger brother $250 a year plus room and board to keep the firm's books and be his assistant.

Field's outgoing personality soon freed him from the drudgery of bookkeeping, and he was sent on the road to sell paper to wholesalers in New York and New England, a task at which he excelled. On his travels he met Mary Stone, whom he had known as a child, and fell in love. They married in 1840, when he was barely twenty-one years old, and the marriage proved an enduringly happy one. And, because she wanted to live in New York, Field accepted a junior partnership with the paper wholesaling firm of E. Root and Company in the city.

E. Root and Company was not well run and, badly damaged by the crash of 1837 and the ensuing depression, it collapsed in bankruptcy soon after Field joined it. Field, however, managed to settle the old firm's debt for thirty cents on the dollar and acquire its stock-in-trade. He was soon in business for himself under the name of Cyrus W. Field and Company. Before long he was one of the leading paper and printing supplies wholesalers in the country.

By the end of the 1840s Field had a net worth of more than $200,000, according to Moses Y. Beach, editor of the New York *Sun*, who had begun publishing a list of the richest New Yorkers (the precursor of the "Forbes 400" list) in 1845. In recognition of his success, he built a large house uptown on the corner of Lexington Avenue and Twenty-first Street, facing fashionable Gramercy Park. His brother David built one next door. They would both live there for the rest of their lives.

Together they hired the cabinetmaker Charles A. Baudouine, a New Yorker of French Huguenot descent, to decorate their houses. This is said to have been the first time a professional decorator was employed in New York for private houses, and Baudouine did them up in the overstuffed high Victorian style then approaching the peak of its popularity. When he was done,

"Louis XIV furniture abounded, as did Italian draperies, Greek Statues, marble mantels, and frescoed ceilings."[12]

Many of their neighbors were among the city's most notable citizens, including Samuel J. Tilden, the Democratic nominee for president in 1876; James Harper, head of Harper and Brothers Publishers, the country's largest publishing house, and onetime mayor of New York; and George Templeton Strong, a successful lawyer, whose vast diary, running from 1835 to 1875, is an indispensable and often highly entertaining resource for the history of his time and place.

Cyrus also began traveling extensively, an activity for which he would never lose his taste, despite chronic susceptibility to seasickness. But even when on "vacation," his relentless drive was always evident. Returning from a grand tour of Europe, his daughter remembered that "Mr. Field amused his friends by stating the characteristic fact that the first word he learned of each new language, as he crossed from one country to another, was, 'faster.'"[13]

Field's business continued to expand in both size and profitability. In 1852 it had gross sales of more than $800,000, well over twice what it had done five years earlier. Field then did a most unusual thing. Although under no obligation to do so, he paid off the debts, with interest, of E. Root and Company, which he had settled, and been released from, for thirty cents on the dollar ten years earlier. One surprised recipient wrote Field that he wished to express "our thanks for the sum enclosed, not so much for the value thereof in currency as for the proof it affords that 'honesty still dwells among men.'"[14] There is no doubt that whatever this gesture cost him in money, it was far less than what he gained in reputation.

Like all natural-born entrepreneurs, Field, only in his early thirties, was not content simply to run a successful and growing business. He needed new worlds to conquer. He turned over management of the firm to his brother-in-law and partner, Joseph Stone, and set off with his friend, the great landscape painter Frederic Church, to explore South America.

Arriving at Barranquilla, Colombia, Field and Church struck out overland to Bogotá on muleback. They stayed for a while in the Bogotá area so Church could do sketches nearby for one of his most famous paintings, *The Falls of Tequendama*. Dozens of workmen were employed to fell trees from an area the size of a city block so that Church could have the view he desired. On July 4, the two Americans rounded up all the English-speaking people they could find in Bogotá—most of whom were, in fact, British—and gave a banquet.

Then the two set out on muleback for Guayaquil, the main port of Ecuador on the Pacific Ocean, more than six hundred miles away. The trail, such as it was, crossed the mountain wilderness of the northern Andes, ranging from tropical rainforest to well above the snow line. The trip took four months. Church sketched endlessly while Field collected biological specimens. Sometimes they were entertained handsomely at haciendas; at other times they spent the night in filthy mud huts.

Finally arriving in Guayaquil, they set off for Panama by ship, crossed the Isthmus of Panama, and returned to New York by steamer. The arrival of Field and Church in New York, on October 29, 1853, created a minor sensation, as Field had in tow, besides many specimens and souvenirs, twenty parrots and parakeets, a jaguar on a leash, and the son of one of his guides, a fourteen-year-old boy named Marcos, who moved in with the family on Gramercy Park. The boy, whom Field hoped to educate, caused endless trouble. "A civilized life was not attractive to him," Field's daughter noted laconically in her biography of her father.[15]

A few years later, when Field was away in England, the family summarily shipped him back to his father, a bullfighter, in Colombia.

Back from his great adventure, semiretired from business, and only thirty-four years old in November of that year, Cyrus Field was at loose ends, undecided what to do next. As so often happens in Victorian novels, fate promptly provided an answer when his brother Matthew brought a Canadian engineer named Frederick Gisborne to see him.

III

NEWFOUNDLAND

MATTHEW FIELD HAD LEFT the paper business and become a successful civil engineer. At this time he was staying at the Astor House, New York's best-known hotel, located across from the foot of City Hall Park on Broadway, when he met Gisborne, who was also staying there.

Frederick Gisborne had been born in England but immigrated to Canada as a young man. He farmed in Quebec for a couple of years before becoming the chief operator for the Montreal Telegraph Company just as the new technology began to sweep the world. He was soon working for the Nova Scotia Telegraph Company, building telegraph lines connecting Montreal with the Maritime Provinces of New Brunswick and Nova Scotia, becoming a largely self-taught engineer in the process. To the north and east of these British colonies, as they then were, across the Gulf of St. Lawrence, lay another crown colony, the rugged, remote, and often fogbound island of Newfoundland.*

Although fishermen had been settled in Newfoundland for centuries, the island's population was only a few tens of thousands

---

*The self-governing Dominion of Canada came into being in 1867, consisting of Ontario, Quebec, Prince Edward Island, New Brunswick, and Nova Scotia. Newfoundland remained a crown colony, however, until it joined Canada in 1949 as its tenth province.

in the mid–nineteenth century and was largely restricted to iso-
lated fishing villages along its deeply indented eastern and southern coastline. Communications between these villages was almost
entirely by sea, as the island was nearly devoid of roads. At forty-
two thousand square miles, Newfoundland is the fourth-largest
island in the Atlantic Ocean, a third again as large as the island of
Ireland. But its vast interior was still largely unexplored in the
middle of the nineteenth century and remains largely uninhabited
to this day.

It would seem an unlikely place to attract interest in the new
technology of telegraphy. But besides its proximity to the great
fishing grounds that have been the island's lifeblood since
Europeans first settled there, Newfoundland possessed one other
great advantage: it is the closet point in North America to
Europe. St. John's, the island's capital and only city, located near
the southeast corner of the island, is almost twelve hundred
miles — one-third of the total distance — closer to the British Isles
than New York, and lies close to the shortest route as well.

The first to notice the possibilities for telegraphy in this geo-
graphical propinquity was the Roman Catholic bishop of
Newfoundland, J. T. Bullock. On November 8, 1850, Bishop
Bullock wrote a letter to the local newspaper in St. John's pointing it out.

Sir:

"I regret to find that, in every plan for transatlantic communication,
Halifax [the capital of Nova Scotia] is always mentioned, and the natural capabilities of Newfoundland entirely overlooked. This has been
deeply impressed on my mind by the communication I read in your
paper of Saturday last, regarding telegraphic communication between
England and Ireland, in which it is said that the nearest telegraphic station on the American side is Halifax, twenty-one hundred and fifty-five

miles from the west of Ireland. Now would it not be well to call the attention of England and America to the extraordinary capabilities of St. John's, as the nearest telegraphic point? It is an Atlantic port lying, I may say, in the track of the ocean steamers, and by establishing it as the American telegraphic station, news could be communicated to the whole American continent forty-eight hours, *at least*, sooner than by any other route.

The Bishop suggested running a telegraph line from St. John's to Cape Ray, at the southwest tip of Newfoundland. From there, he proposed a submarine cable across the Cabot Strait to the northern tip of Cape Breton Island, the northern part of Nova Scotia.

"Thus, it is not only practicable to bring America two days nearer to Europe by this route, but should the telegraphic communication between England and Ireland, sixty-two miles, be realized, it presents not the least difficulty.

"Of course, we in Newfoundland will have nothing to do with the erection, working, and maintenance of the telegraph; but I suppose our government will give every facility to the company, either English or American, who will undertake it, as it will be an incalculable advantage to this country. I hope the day is not far distant when St. John's will be the first link in the electric chain which will unite the Old World and the New."[1]

At the time Bishop Bullock published his letter, there were as yet no working submarine cables at all. The Brett brothers would not lay the successful English Channel cable until the following year. But Gisborne, far more familiar with the technical aspects of telegraphy, was already thinking along lines similar to the bishop's, though his proposal was somewhat more conservative. The following spring he proposed to the Newfoundland legislature that a 400-mile telegraph line be built from St. John's to Cape Ray and that it be connected to the telegraph on Cape Breton by means of steamers and carrier pigeons. The legislature voted to appropriate £500 to fund an exploratory survey.

Gisborne resigned as head of the Nova Scotia Telegraph Company and devoted himself to the new project. The survey, begun in September 1851, proved a difficult business. The south coast of Newfoundland is extremely rugged and to this day there is no road along it; the Trans-Canada Highway to St. John's lies far inland. After the survey was finally finished, Gisborne wrote in a letter that "on the fourth day of December, I accomplished the survey through three hundred and fifty miles of woods and wilderness. It was an arduous undertaking. My original party, consisting of six white men, were exchanged for four Indians; of the latter party, two deserted, one died a few days after my return, and the other, 'Joe Paul,' has ever since proclaimed himself an ailing man."[2]

Gisborne went to New York and arranged for backing from a syndicate of investors. Returning to St. John's in the spring of 1852, he obtained a charter from the legislature giving his Newfoundland Electric Telegraph Company exclusive rights for thirty years to build telegraphic lines on the island and granting the company considerable land upon the successful completion of the line to Cape Ray.

Gisborne then went to England to arrange for supplies, and there he consulted with the Brett brothers regarding undersea cables. On his return he laid the first true submarine cable in the Western Hemisphere, running nine miles under the Northumberland Strait, which separates Prince Edward Island from New Brunswick.

He soon turned to completing the line across Newfoundland, with a crew of 350 men. But it turned out to be much more difficult than the initial survey had indicated. He had originally intended to lay most of it underground but quickly abandoned the attempt when the soil in many areas turned out to be only inches deep. Indeed, he found it impossible even to dig holes for poles in many areas and the poles had to be supported by piles of rock, perhaps the one land resource that the south coast of Newfoundland has in limitless abundance.

Before he had gone more than forty miles—about one-tenth of the total distance—he was out of money and his partners in New York refused to provide more, despite their promises. As Gisborne's was the only name to appear on the charter of incorporation, he was held wholly responsible. He was arrested and whatever assets he had in Newfoundland were seized. Gisborne, an honorable man, did what he could to pay the company creditors, mostly workmen, but his company was bankrupt before the project was fairly begun. He made his way to New York in hopes of finding new funding.

Wholly by chance he met Matthew Field at the Astor House, and Field managed to talk his initially reluctant brother Cyrus into hearing what Gisborne had to say.

Like most good businessmen, Cyrus Field was a good listener, and he listened quietly as Gisborne outlined his idea in Field's library one evening. But Field was not much impressed with the idea. It was true that St. John's was twelve hundred miles along the route to Europe from New York. Thus, at the current speed of steamers at that time, roughly ten knots, a telegraph line from St. John's to Cape Ray and then across the Cabot Strait would speed up the transmission of news from Europe somewhat.

But Halifax, a far larger and more important city than St. John's (it is Canada's only ice-free mainland port on the Atlantic), was six hundred miles en route to Europe and was already connected to the burgeoning telegraph network. Field could see little reason to go to the trouble, expense, and risk of building a telegraph line across an inhospitable wilderness and laying a considerable submarine cable, just to shorten news travel by a day.

But after his brother and Gisborne left, Field looked at the globe in his library. He suddenly realized that if a submarine cable could be laid across the Atlantic from Ireland to Newfoundland, the time needed for news to cross the ocean would be shortened not by a couple of days but by a couple of weeks. That was a different matter altogether.

London was the center of the financial world and Britain the world's foremost economic power in the 1850s. The United States, though growing explosively, was still a major importer of capital to fund construction of its railroads and industries, and that capital came mainly from Britain. Furthermore, Britain was then truly the "workshop of the world," producing the majority of the world's industrial goods. Thus, what was happening in "the City," as London's financial district is called, was of the utmost importance to every businessman in America, and, as that very successful American businessman Benjamin Franklin had noted a century earlier, "time is money."

Field knew that to own the means of instant communication with the London markets would be to own an instrument of potentially formidable profit. But could it be done? Could someone, in Henry Field's words, "lay a bridle on the neck of the sea"?3

Field knew little of the science or technology involved in such an undertaking. Indeed, it is probably fortunate that he did not,

Samuel F. B. Morse

because he might well have dismissed the entire notion as impossibly fanciful. But the next day he wrote to two experts for their advice. One was Samuel F. B. Morse.

Morse, like Field, was the son of a notable New England Congregational minister, Jedidiah Morse, who was also the founder of American geography. His first book, *Geography Made Easy*, published in 1784, was the first work on the subject published in the United States. Samuel Morse was a generation older than Field, having been born in 1791. He studied to be an artist and had a moderately successful career as a portrait painter. His larger historical works, however, were not very successful, and he began to turn his attention to telegraphy. Between 1832 and 1844 he struggled to perfect his system, often incorporating the ideas of others, such as Joseph Henry, the first head of the Smithsonian Institution and perhaps the most distinguished scientist in the United States at that time, and Alfred Vail, his business partner.

After success was finally achieved with a government-financed line between Baltimore and Washington, D.C., he had to defend his patents vigorously in court and did not finally succeed in establishing his priority until 1854, when the U.S. Supreme Court decided in his favor. Once his patents were secure, however, they made Morse a very rich man.

Even before he achieved success on land in 1844, Morse was confident that an Atlantic cable was possible. "A telegraphic communication on the electro-magnetic plan may with certainty be established across the Atlantic," he had written Secretary of the Treasury John C. Spencer in 1843, relating his experiments with underwater telegraph lines across New York Bay. "Startling as this may now seem, I am confident that the time will come when this project will be realized."[4] He told Field that his confidence was no less now than it had been a decade before and offered his enthusiastic support. Field, almost wholly lacking in technical knowledge himself, probably did not realize how limited was Morse's own knowledge of the technical problems involved.

The other expert to whom Cyrus Field wrote for advice was Lieutenant Matthew Fontaine Maury of the United States Navy. Maury, a Virginian, had joined the navy in 1825 at the age of nineteen. He spent four years circumnavigating the globe on the USS *Vincennes,* and remained on sea duty until 1834, when he took a leave of absence to write *A New Theoretical and Practical Treatise on Navigation.* Increasingly interested in improving navigation, he was assigned to surveying the southeast coast of the United States. In his spare time he published numerous articles, lobbying for increased expertise and professionalism in the navy. Widely read, these articles were important in persuading Congress to establish the United States Naval Academy at Annapolis, Maryland, in 1845.

A stagecoach accident in 1839 rendered Maury permanently lame and he did not serve again at sea. He was named head of the Depot of Charts and Instruments, and in 1844 the navy appointed him head of the new United States Naval Observatory in Washington. He devoted the rest of his naval career to charting the winds and currents of the world's oceans and promoting international data gathering for meteorology.* He represented the United States at the Brussels International Meteorological Conference in 1853, where uniform standards for meteorological data were developed. By using the new knowledge that was becoming available through his efforts, Maury charted a better route between New York and San Francisco and reduced the average time needed for the trip by no less than forty-seven days.

---

*His career ended in 1861, when he resigned his commission after his native Virginia seceded from the Union and offered his services to the Confederacy.

Lieutenant Matthew Fontaine Maury

By the early 1850s, Maury was the nation's premier oceanographer, and in 1855 his *The Physical Geography of the Sea*, the first American textbook on oceanography, made him internationally famous and proved to be one of the great scientific best-sellers of the nineteenth century.

It was only natural, therefore, that Cyrus Field would turn to Maury to learn if laying a cable across the Atlantic Ocean was a practical possibility from an oceanographic standpoint. Maury promptly replied, writing that "singularly enough, just as I received your letter, I was closing one to the Secretary of the Navy on the same subject." He enclosed a copy.

Maury had written the secretary that the United States brig

Dolphin had spent the summer of 1853 surveying the winds and currents of "that part of the ocean which the merchantmen, as they pass to and fro upon the business of trade between Europe and the United States, use as their great thoroughfare." The *Dolphin*'s captain had taken the opportunity to make a line of deep sea soundings between Newfoundland and Ireland, providing precisely the information Field needed. Maury wrote,

> This line of deep-sea soundings seems to be decisive of the question of the practicability of a submarine telegraph between the two continents, *in so far as the bottom of the deep sea is concerned.* From Newfoundland to Ireland, the distance between the nearest points is about sixteen hundred miles; and the bottom of the sea between the two places is a plateau, which seems to have been placed there especially for the purpose of holding the wires of a submarine telegraph, and of keeping them out of harm's way. It is neither too deep nor too shallow; yet it is so deep that the wires but once landed, will remain for ever beyond the reach of vessels' anchors, icebergs, and drifts of any kind, and so shallow, that the wires may be readily lodged upon the bottom.
>
> The depth of this plateau is quite regular, gradually increasing from the shores of Newfoundland to the depth of from fifteen hundred to two thousand fathoms, as you approach the other side. . . .
>
> I [do not] pretend to consider the question as to the possibility of finding *a time calm enough, the sea smooth enough, a wire long enough, a ship big enough,* to lay a coil of wire sixteen hundred miles in length; though I have no fear but that the enterprise and ingenuity of the age, whenever called on with these problems, will be ready with a satisfactory and practical solution of them."5

Maury also told the navy secretary that samples brought up from what he called "that beautiful plateau" were entirely of microscopic shells, with no sand or gravel. These shells were the remains of plankton that had drifted down after death, indicating that there were no scouring currents at the bottom that might disturb a submarine cable. "Therefore," Maury concluded, "so far as the bottom of the deep sea between Newfoundland . . . and Ireland is concerned, the practicality of a submarine telegraph across the Atlantic is proved."[6]

Cyrus Field could hardly have received better news. Not only had the obvious route for a submarine cable been explored, it had been found to be nearly ideal for the purpose.

Field decided to see if he could raise enough capital to make it worthwhile to organize a company to take over Gisborne's franchise in Newfoundland and then to lay a cable across the Atlantic. The first person he called on was his Gramercy Park neighbor Peter Cooper.

Cooper, the child of an unsuccessful father, had had little formal education, but he was a born tinkerer and from an early age had helped his father in various enterprises. He had also helped his mother with the household chores, and while still a child, he had invented a machine for pounding dirty clothes, perhaps the world's first washing machine, and designed a mechanical lawn mower at a time when lawn mowers were called sheep.

Frugal by nature (he regularly settled all his debts every Saturday throughout his life), Cooper as a young man managed to accumulate $2,000 and buy a run-down glue factory in New York. A notably good chemist as well as an engineer, he quickly improved the product line and in his first year cleared a profit of $10,000, five times what he had paid for the entire factory. Soon he was making $100,000 a year from his glue factory, a vast income at the time.

He moved into the iron business and produced the very first locomotive ever built in the United States, for the Baltimore & Ohio Railroad, which had previously used horses. The Tom

Thumb, as it was known to history,* was a work of improvisation. Cooper had to take into account the tight curves of the B&O, and he cobbled together some old wheels, a carriage, and a steam engine he had built earlier for another purpose. The only piping available at that time was made of lead, which could not stand up to the heat and pressure of steam, so Cooper sawed off a couple of rifle barrels and used them for piping. To everyone's surprise, except Cooper's, the Tom Thumb functioned well.

By the 1850s Peter Cooper was among the richest men in the United States, but he was not much interested in new enterprises. He was in his sixties and very actively engaged in founding the Cooper Union, a technical college located at Astor Place in Manhattan. Cooper had always regretted his own lack of formal education and wanted to make it possible for people who had to work for a living to get an education by providing evening classes and charging no tuition. The Cooper Union to this day is the only major private college in the United States that is tuition free.

In addition, though Peter Cooper lacked even a vestige of a sense of humor, he had a deep sense of the moral obligations imposed by wealth and a streak of mysticism that made the idea of an Atlantic cable attractive. "It was an enterprise that struck me very forcibly the moment he mentioned it," he wrote later.

> I thought I saw in it, if it was possible, a means by which we would communicate between the two continents, and send knowledge broadcast over all parts of the world. It seemed to strike me as though it were the consummation of that great prophecy, that "knowledge shall cover

---

*It acquired its name only later, after the showman P. T. Barnum had made the midget "General Tom Thumb" an international celebrity. It may be seen today at the Baltimore & Ohio Railroad Museum in Baltimore, Maryland.

the earth, as the waters cover the deep," and with that feeling I joined
him . . . in what then appeared to most men a wild and visionary
scheme; a scheme that fitted those who engaged in it for an asylum. . . .
But believing, as I did, that it offered the possibility of a mighty power
for the good of the world, I embarked on it.7

Cooper, of course, was no fool. He made his support contingent on Field's getting others to agree to come into the enterprise
as well. Field then went to see another New Yorker, Moses
Taylor, who had made a vast fortune from the opportunities presented by the industrial revolution.

Taylor, fifteen years younger than Cooper and thirteen years
older than Field, was the son of John Jacob Astor's business manager. As a young man he went to work for a company that spe-
cialized in trading with Latin America and soon founded his own
importing firm. By the 1850s Moses Taylor and Company was
paying more import duties than any firm in the country except A.
T. Stewart's store. Taylor became a director of the City Bank, an
ancestor of today's giant Citibank, when not yet thirty, and
would be the bank's president for the last twenty-six years of his
life. He also invested in insurance companies, railroads, coal companies, and, especially, gas lighting companies. By the end of the
Civil War, Taylor would control the entire gas lighting industry
in New York City.

Cyrus Field did not know Moses Taylor personally, but his
brother David did and provided a letter of introduction. Taylor
invited Field to his house and listened to what he had to say. "I
shall never forget how Mr. Taylor received me," he remembered
years later. "He fixed on me his keen eye, as if he would look
through me: and then, sitting down, he listened to me for nearly
an hour without saying a word."8 Taylor, though far less concerned with the moral obligations of wealth than was Peter

Cooper, was quite as capable of seeing the economic possibilities. He, too, agreed to invest as long as enough others could be found. He immediately suggested Marshall O. Roberts, another native New Yorker, who was a major shipowner, running ships on the Hudson River and to Panama, then the shortest route to California. Roberts, too, agreed to participate.

Field then persuaded Chandler White, who, like Field, had made a fortune in the paper business, and whom he had known since the days when he had been working for his brother Matthew, to invest in the new enterprise. Field had been hoping to gather a group of ten investors, but when Field had assembled five, including himself, Peter Cooper suggested that "if ten men can carry out the project, so can five."9

Field agreed and arranged for Cooper, Taylor, Roberts, White, and himself to meet with Gisborne. Also present were Samuel Morse and Field's brother David, who would act as the attorney for the enterprise. They first met at the Clarendon Hotel, located far uptown at Broadway and Thirty-eighth Street. Then, on four subsequent nights, the group assembled in Field's dining room, where maps and charts could be spread out on the table.

Years later Daniel Huntington, a leading American portrait artist, who had studied under Samuel Morse, would paint *The Atlantic Cable Projectors*, which enjoyed wide popularity. Though not strictly accurate, as it included Daniel Huntington himself as well as Wilson G. Hunt, who would not join the group until after the death of Chandler White in 1856, it is still regarded as one of better group portraits of the era.

The group agreed on March 10, 1854, to take over the New-foundland Electric Telegraph Company from Gisborne, assuming its debts of about $50,000, and to form a new company to be called the New York, Newfoundland, and London Telegraph

*Daniel Huntington's painting. Left to right: Peter Cooper, David Dudley Field,
Chandler White, Marshall O. Roberts, Samuel Morse, the artist Huntington,
Moses Taylor, Cyrus Field, and Wilson G. Hunt.*

Company. But all of this was contingent on persuading the government of Newfoundland to grant the new company a more
favorable charter than it had granted to Gisborne. The two Field
brothers and Gisborne agreed to go at once to St. John's and
approach the government there.

It was a miserable trip. They left New York on March 14 by
steamer for Boston, transferred there the next day to one bound
for Halifax, and on March 18 left Halifax for St. John's on the
small steamer *Merlin*. David Field wrote

> Three more disagreeable days, voyagers scarcely ever passed, than we
> spent in that smallest of steamers. It seemed as if all the storms of win-
> ter had been reserved for the first month of spring. A frost-bound coast,
> an icy sea, rain, hail, snow and tempest, were the greetings of the tele-

graph adventurers* in their first movement toward Europe. In the darkest night, through which no man could see the ship's length, with snow filling the air and flying into the eyes of the sailors, with ice in the water, and a heavy sea rolling and moaning about us, the captain felt his way around Cape Race [the southeastern most point of Newfoundland] with his lead [taking soundings] as the blind man feels his way with his staff, but as confidently and as safely as if the sky had been clear and the sea calm; and the light of morning dawned upon the deck and mast and spar, coated with glittering ice, but floating securely between the mountains which form the gates of the harbor of St. John's.[10]

The adventurers quickly disembarked and went to call on Edward M. Archibald, attorney general of Newfoundland. Archibald was delighted with this turn of events and soon introduced the trio to the governor, Kerr Bailey Hamilton. Governor Hamilton convened his council and they agreed to ask the Newfoundland Assembly to grant a charter giving the new company exclusive rights to lay telegraph lines on the island and cables touching Newfoundland for fifty years. Together with this monopoly, the charter also gave a guarantee of the interest on £50,000 worth of bonds, an immediate grant of fifty square miles of land and another fifty upon completion of the cable to Ireland, and a grant of £5,000 toward the completion of a bridle path along the line of the telegraph to Cape Ray.

After debate, the Assembly, anxious to pay off the workmen from the last attempt and hopeful that the project would be an economic boost to Newfoundland, passed the necessary legislation

---

*Field here is using the word *adventurer* in the sense of one who invests in an enterprise, a meaning that came into use in the dawn of capitalism in the early seventeenth century. The obsolete term is echoed today in the phrase *venture capitalist*.

with only a single dissenting vote. Cyrus Field rushed back to
New York and arranged to transfer $50,000 to a Newfoundland
bank to pay the workmen.

The charter was highly favorable to the interests of the New
York, Newfoundland, and London Telegraph Company, which
is not surprising, as it had been written almost in its entirety by
David Dudley Field, during the passage from Halifax. Indeed, it
was so favorable that the government in London later noted that
"Her Majesty will be advised not to give her ratification to the
creation of similar monopolies."[11] It accepted this one, however,
and the charter was celebrated in typical Victorian fashion with a
large banquet.

David Field and Chandler White did not get back to New
York from Newfoundland until the evening of Saturday, May 6.
Because some were unwilling to do business on the Sabbath and
one of the group had to leave the city early Monday, the organizational meeting was held at six in the morning on May 8 at the
house of David Field. Cooper, Cyrus Field, White, Taylor, and
Roberts were named directors. Peter Cooper was elected president, Chandler White vice president, and Moses Taylor treasurer. Cyrus Field was given no formal position, as it was understood that he would be, in effect, chief operating officer of the
enterprise and far too busy to handle details. He was soon to leave
on the first of what turned out to be more than fifty Atlantic
crossings on company business.

The group members committed themselves to raising capital
in the amount of $1.5 million. That was a huge sum for those
days,* but it would prove to be nowhere near enough. It would

---

*In 1854, the total expenditure of the federal government was only a little more
than $58 million.

not have been nearly enough even if everything had gone right, which it certainly didn't.

Thus, five of the most successful and savvy businessmen in New York had, in fifteen minutes before breakfast one bright May morning, casually committed themselves, in terms of both money and personal prestige, to one of the epic undertakings of the nineteenth century. As Cyrus Field ruefully admitted fourteen years later, "God knows that none of us were aware of what we had undertaken to accomplish."[12]

IV

"How Many Months? Let's Say How Many Years!"

The plan was first to accomplish the link between New York and Newfoundland. This would give the company an asset that would generate some revenue and was regarded as the easy part of the whole venture.

It required three steps: completing the line across southern Newfoundland, laying a submarine cable across the Cabot Strait that separated Newfoundland from Cape Breton Island, and building a 140-mile line on Cape Breton Island to connect to the existing telegraph system in Nova Scotia. Chandler White was appointed to run the office in St. John's and Matthew Field was appointed supervising engineer, with Frederick Gisborne as consultant, of the Newfoundland portion.

Gisborne's first company had strung about 40 miles of the line, but it was the easiest 40 miles, nearest St. John's and through the most settled country. The next 360 miles were nearly uninhabited except for the fishing villages found at the head of the fjords that frequently interrupt the coastline, which consists mostly of towering cliffs. Swift-flowing streams rush down from hills of nearly 2,000 feet to these fjords, carrying the rain and summer snowmelt. Dense fog is often a problem on this coast, especially in summer, thanks to the meeting of the Labrador Current and the Gulf Stream over the nearby Grand Banks. At least the south coast of Newfoundland is reliably ice free.

The overland construction party, which numbered about six hundred, could only be supplied by sea, and Cyrus Field purchased a steamer in Halifax, the *Victoria*, to do so. The ship plied continuously between Halifax, St. John's, and whatever fjord was closest to the working party, carrying "barrels of pork and potatoes, kegs of powder, pickaxes and spades and shovels, and all the implements of labor."[1] Once put ashore, these had to be carried up the hills to the camp on the backs of the men. In clear weather, the cookfires of the camp could often be seen far out at sea. But the weather was not often clear. Rain was frequent and heavy in the summer and the snows deep in the winter.

Even by the standards of that time, when people were far more used to hard physical labor than today, the work was both brutal and unrelenting, except when the weather was so foul the men could only huddle in their tents. As Henry Field wrote, "How at such times the expedition lay floundering in the woods, still struggling to force its way onward; what hardships and sufferings the men endured—all this is a chapter in the History of the Telegraph which has not been written, and which can never be fully told."[2]

Peter Cooper's wire factory in Trenton, New Jersey, supplied the construction team with telegraph wire, but the demands for money from Matthew were unceasing. It had been lightly thought that the Newfoundland phase of construction could be completed in the summer of 1854. In fact, it would take well over a year to finish and cost $500,000, one-third of the firm's capital.

As Matthew Field worked his way slowly across the Newfoundland wilderness, Cyrus Field went to England to arrange for the manufacture of cable to cross the Cabot Strait. While Peter Cooper's firm could produce excellent telegraph wire for landlines, only Britain had the capacity at that time to produce submarine telegraph cable. The reason was not so much Britain's technological and industrial lead over the rest of the world, considerable as that was in the 1850s, as its monopoly on a now nearly forgotten material, gutta-percha.

Humans have used any number of natural materials—stone, wood, and plant fibers—for an infinity of purposes since long before recorded history. But after the discovery of metals around 3500 B.C., few new substances came into use until very recent times. The eighteenth century utilized virtually no materials unknown to the Romans except porcelain. And the twentieth century would prove to be a golden age of new materials, with the explosive development of plastics, artificial fibers, and ceramics.

But the chemical foundations of the new science of materials that has transformed the world in recent decades were, as so much else, laid down in the Victorian era. Coal tar—the residue left after coal is heated in the absence of air to produce the coal gas that lighted the Victorian urban world—proved to be a rich source of new substances. Among them were naphtha, which Charles Mackintosh used in 1820 to produce the first waterproof cloth, and creosote, which preserved wood, most significantly the railroad ties and telegraph poles that were knitting the world economy together. Aniline dyes, first synthesized from coal tar by the Englishman William H. Perkin in 1856, made a rainbow of colors for cloth, paper, and paint available for the first time.*

Rubber was another substance whose chemistry was explored

---

*These new dyes also freed up millions of acres of agricultural land, once used for such dye crops as madder (which produced a red dye) and indigo (which produced a blue one) for food production. This is one of many reasons that Thomas Malthus was wrong and why food prices fell in the nineteenth century while population rose swiftly.

in the early nineteenth century, notably by the American Charles Goodyear, who discovered in 1839 that rubber, when mixed with sulfur and heated, became stable, its plasticity no longer a function of temperature. The new material would prove to have any number of industrial and commercial uses, notably tires, but also bushings, gaskets, and belts in machinery.

For submarine telegraphy, however, none of the new substances were nearly as important as gutta-percha. It comes from several trees native to Malaya, mostly of the genus *Palaquium*. It is chemically similar to rubber, and both are natural plastics, polymers in chemical terms. But gutta-percha is only moderately flexible, not elastic like rubber. Also unlike rubber, it does not deteriorate when immersed in water for long periods.

The gray, milky sap of the gutta tree is boiled to produce gutta-percha, which is naturally gray in color. Because the latex does not flow copiously when tapped, as the sap of the rubber tree does, the trees must be killed in the process of obtaining it. Soft and pliable at temperatures above one hundred degrees Fahrenheit, at room temperature gutta-percha becomes hard (a fingernail can barely mark it) while remaining flexible.

Gutta-percha had been known for centuries to the Malayans, who used it to make whips, tubs, knife handles, and other objects. It was first brought to Europe, as a curiosity, in the seventeenth century, and it remained just that until the 1840s. In that decade, however, the boom in "economic botany" reached a high point, with European and American botanists scouring the world for plants and plant materials that might have commercial applications. Many of today's most treasured garden flowers—many varieties of rhododendron and primroses from China, for instance—first appeared in the west at this time.

In 1843 a Portuguese engineer named José d'Almeida described gutta-percha and its uses to the Royal Asiatic Society in London. A few months later an East India Company surgeon wrote a paper on gutta-percha and sent samples to London that were displayed at the Society of Arts. The stuff immediately became a craze. Soon British rubber manufacturers were using gutta-percha to make such items as boot soles, bottle stoppers, snuffboxes, objets d'art, jewelry, and walking sticks. It would also revolutionize the game of golf, when a golf-playing Scottish clergyman made golf balls out of gutta-percha. They proved both much cheaper and longer-lasting than the old "featheries" made of leather stuffed with boiled goose feathers and soon replaced them.*

Werner von Siemens in Germany, founder of the great electronics firm that bears his name, soon determined that gutta-percha was a good electrical insulator (as, indeed, nearly all plastics are). He quickly developed a press, not unlike a macaroni-making machine, that extruded a sheath of gutta-percha around a copper wire. Michael Faraday, one of the giants of nineteenth-century science, also investigated gutta-percha and wrote a short paper on gutta-percha's electrical qualitites in 1848.† It was soon the preferred material for waterproof insulation and a vast improvement on tarred hemp, which was all that had been available before. The Gutta Percha Company, headed by Willoughby Smith, quickly established a near monopoly on the importation of

---

*Gutta-percha golf balls turned out to have one other, highly appealing virtue. It was quickly noticed that the new balls could be driven farther after they had been used a few times. The reason turned out to be that the dents and nicks they acquired in use gave them better aerodynamics. Equipped with the dimples that have been a feature of golf balls ever since, the new balls could be driven on average about twenty-five yards farther than the old featheries.

†Among numerous other accomplishments, Faraday proved the identity of electricity and magnetism in the 1830s, the first step toward the unified field theory of physical forces that is still the Holy Grail of modern physics.

the substance from Malaya, and worked hard to develop its electrical possibilities.

In 1850 the Brett brothers ordered twenty-five miles of cable insulated with gutta-percha for their English Channel project. After their success the following year, gutta-percha became standard for the manufacture of submarine cable and demand for the material soared. But because each gutta tree yields only two or three pounds of gutta-percha, and long submarine cables could require hundreds, even thousands, of tons, the next fifty years would see the gutta trees extirpated from much of their native range.

To order the cable, Field sailed for England near the end of 1854. He had been delayed, first because of the death of his brother-in-law, Joseph Stone, who was also his partner in the paper business. Field had no choice but to resume running Cyrus Field and Company once more, at least for a while.

Then in August, his only son, Arthur Stone Field, age four, died suddenly, plunging his father into grief. In the 1850s, more than half the children born in New York City died before reaching the age of five. Often the cause was one of such waterborne diseases as cholera and typhoid that were only beginning to ebb as clean water and proper sewage removal spread through the city. However, fresh milk—now brought in from the countryside by the new railroads at affordable prices—proved to be a new and potent source of infant mortality. Every summer thousands of children would come down with "summer complaint," a violent diarrhea that we now know was caused by various microorganisms that thrived in milk. Many of these children did not survive. Infant mortality was, of course, more common in the fast-spreading slums, but even in the most affluent households, such as that of Cyrus Field, it was tragically common for a seemingly healthy child suddenly to sicken and die.

He did not reach England until the new year and immediately went to see John Brett, who now headed the Magnetic Telegraph Company in England and was the acknowledged expert on submarine telegraphy. While many in England laughed at Field's project, Brett did not. Indeed, he was the first Britisher to invest in Field's company.

Brett took Field around to the various cable manufacturers and suggested that Field choose a cable consisting of three copper wires, each individually insulated with gutta-percha and then bundled together, wrapped in tarred hemp and another layer of gutta-percha and the whole sheathed in galvanized iron wire. Field, who had no expertise whatever in these matters, had no choice but to accept Brett's recommendation and placed the order with Küper and Company. He arranged to have the completed cable loaded aboard a 500-ton brig named the *Sarah L. Bryant*, ready for laying. She was to make for Port aux Basques,* near Cape Ray, the Newfoundland terminus of the cable across the Cabot Strait. He also hired Samuel Canning, a young British engineer who had experience laying cables in the Mediterranean, to supervise the project.

BACK IN NEW YORK by March 1855, Field wrote to his brother Matthew asking how long it would be before completion of the landline from St. John's to Cape Ray. Matthew was straightforward, a characteristic of the Field brothers. "How many months?" he asked rhetorically. "Let's say how many *years!* Recently, in building half a mile of road we had to bridge three

---

*Basque fishermen had been catching cod in Newfoundland waters for decades before Columbus "discovered" America.

John Brett

ravines. Why didn't we go around the ravines? Because Mr. Gisborne had explored twenty miles in both directions and found more ravines. That's why! You have no idea of the problem we face. We hope to finish the land line in '55, but wouldn't bet on it before '56, if I were you."[3]

But while the slowness of the landline's progress was disappointing, there was no reason to delay laying the submarine cable on account of it. When the cable, all eighty-five miles of it, was finished and on its way, Field chartered the *James Adger*, a coastal steamer that regularly ran between Charleston, South Carolina, and New York, at the cost of $750 a day. The ship was needed to tow the *Sarah L. Bryant* across the strait while the latter vessel laid the cable, but she also provided luxurious accommodation

for the directors, wives, children, and assorted friends and distinguished guests.

Among those invited to take part in the expedition who
accepted were Mr. and Mrs. Peter Cooper; Mr. and Mrs. Samuel
Morse and their son; the seventy-four-year-old Reverend David
Dudley Field; the equally prominent Reverend Gardner Spring,
rector of the Brick Presbyterian Church in New York for an
astonishing sixty-three years; Henry Field and his wife; and two of
Cyrus Field's daughters. Cyrus Field's wife, Mary, however, was
not along. She had given birth the previous month to a second son,
Edward Morse Field (his middle name honored Samuel F. B.
Morse) and was still nursing him. Altogether the party numbered
more than fifty people, including several reporters from New York
newspapers.

On the morning of August 7, 1855, a beautiful day, the *James
Adger* departed New York on what was, in about equal measure,
a yachting party, a cutting-edge technological expedition, and a
capital-intensive business venture, bound for Halifax and then
Newfoundland. On deck, Samuel Morse demonstrated telegraphy with a portable apparatus he had brought along; Peter
Cooper and the two distinguished clergymen discussed theology;
sailors strummed guitars to entertain the ladies; and the ship
plowed northward through initially perfect weather that soon
turned foul. Several passengers, including Henry Field, had to
disembark at Halifax and the ship then made for Port aux
Basques, hoping to find the *Sarah L. Bryant* there, ready to lay the
cable.

The little harbor proved to be empty of shipping, and Samuel
Canning rowed out to the *James Adger* to say that there had been
no sign of the *Sarah L. Bryant*, which had evidently been delayed
by weather. Field decided to go to St. John's, hoping, no doubt,

to sight the other ship en route. At the pier at St. John's the expedition was greeted by a pack of tail-wagging Newfoundlands, the large, black, intelligent, long-furred dogs, "wolf-colored and web-footed,"[4] for which the island has long been known. Dozens of these dogs wandered the city, and the passengers promptly bought several of them to take home.

Always conscious of public relations, Field invited the city council to a party on board and the passengers, in turn, were invited to a ball ashore.

Back at Port aux Basques, the *Sarah L. Bryant* had appeared, in need of repair after a rough passage and with her cable-laying equipment still unassembled. Field and Samuel Canning, reconnoitering by rowboat, decided to land the cable at Cape Ray cove, which had a smooth, sandy bottom, a few miles west of Port aux Basques. The *Victoria* brought lumber from Port aux Basques to the cove to build a shelter for the telegraphic crew that would be stationed there, but the only way to get the lumber ashore was to float it in as rafts, each raft guided by a sailor. As the rafts approached the shore, however, the surf broke the rafts apart, flinging the men into the sea and covering the water with planks, beams, and packages of shingles.

The dogs aboard the *James Adger* immediately came to the rescue. As Bayard Taylor, a reporter for the *New York Tribune*, described it, "the dogs rendered capital service—plunging boldly into the sea and seizing upon every stick which they could manage. Sometimes two of them would take a plank between them, and watching the proper moment with a truly human sagacity, bring it to the beach on the top of a breaker and there deliver it into the hands of their masters."[5]

Once the shelter was built and the *Sarah L. Bryant* repaired, fog immediately closed in and nothing could be done for two days. And when the weather cleared up, nearly everything else went wrong. As the *James Adger* took the *Sarah L. Bryant* in tow, the sailing vessel discovered that her anchor was foul and had to be slipped. The *Adger* began drifting down upon the *Sarah L. Bryant* and hit her broadside to bow, carrying away several shrouds. In response, Captain Turner of the *Adger* cut the tow rope and steamed off, leaving the *Sarah L. Bryant* to fend for herself. The crew tried to lower her remaining anchor but lost it and had to cut the telegraph cable, hurriedly hoist sail, and claw out to sea before she ran aground.

Several people accused Captain Turner of deliberately ramming the *Sarah L. Bryant*, and there is no doubt that he was not in a cooperative mood. He had taken umbrage at the fact that Field had placed the Reverend Spring at the head of the table, which the captain felt belonged to him. And Turner also resented taking orders from a mere passenger, although that passenger, Cyrus Field, was paying $750 a day for the use of the steamer, the crew, and the captain's services.

When the *Sarah L. Bryant* had been repaired and her anchors retrieved, the cable laying began again. And again Turner did his considerable best to frustrate success. A flag had been placed on top of the shelter built the previous day and the captain was told to steer a course that would keep the flag in line with a white rock on the mountainside behind. This line, it was calculated, was the shortest course to the landing site on Cape Breton Island.

The captain, however, proceeded to steer his own course, far out of line with the one indicated, until Peter Cooper went up to him and demanded that he follow instructions.

"I know how to steer my ship," he replied, "I steer by my compass."[6]

Cooper could see how far off the shortest course they were and again demanded that the captain follow his instructions, warning him that he would be responsible for all losses otherwise. The captain ignored him, so Cooper instructed Robert W. Lowber, a company lawyer on board, to draw up a document stating the captain's responsibility. The captain's response was to change course, but only to steer equally far off course in the other direction. The abrupt change in course caused a kink to form in the cable, and the erratic steering meant that while the *Sarah L. Bryant* had payed out twenty-five miles of cable, the ships were only nine miles from shore. At that rate, there would be only enough cable to reach two-thirds of the way across the Cabot Strait.

But besides ignoring his instructions with regard to course, the captain also refused to adjust his vessel's speed to take into account the speed at which the paying-out machinery could pass the cable from the hold of the *Sarah L. Bryant* to the sea. As the captain of the *James Adger* thrashed along, the cable acted more and more like an anchor, increasingly dragging down the stern of the helpless *Sarah L. Bryant*.

Meanwhile, the breeze had been freshening into a gale and the seas were rising, never an easy condition for a vessel under tow to deal with, even when not attached to tons of submarine cable at its stern. Finally, Samuel Canning, aboard the *Sarah L. Bryant*, was forced to cut the cable before his ship was dragged under by it. The laying of the first submarine cable through open waters in the Western Hemisphere, and the first link of a telegraph cable to Europe, had been an ignominious failure.

Cyrus Field was crushed. But, as he ordered both vessels to proceed to Sydney, the largest port on Cape Breton Island, he managed to have Samuel Canning transferred to the *James Adger* from the *Sarah L. Bryant*. Before the assembled passengers, he told Canning that he was in no way responsible for the failure, but rather that he had done his best under impossible circumstances. Canning was deeply grateful for this vote of confidence from a man who might well have been looking for scapegoats. The *Sarah L. Bryant* unloaded her remaining cable in Sydney and the *James Adger* made for New York. Despite the failure, the atmosphere of a yachting party still prevailed and the passengers held a costume party the last night aboard.

THE DIRECTORS HAD MUCH TO DISCUSS when they met in New York. The cable-laying expedition had been a total failure, but they had learned valuable lessons. One was that submarine cables could not be laid from sailing ships or vessels under tow, unable to adjust their speed easily and quickly as circumstances required. The cutting-edge technology of telegraph cables required the cutting-edge technology of steam to lay them. The directors also had no trouble assigning personal blame for the failure. As Peter Cooper wrote years later, "We had spent so much money, and lost so much time, that it was very vexatious to have our enterprise defeated by the stupidity and obstinacy of one man." Although a devout Christian, Cooper must have felt some satisfaction in reporting Captain Turner's fate. "This man was one of the rebels that fired the first guns on Fort Sumter. The poor fellow is now dead."7

Cooper persuaded his fellow directors to press on. But, as always, it was Cyrus Field who supplied most of the drive. Cooper wrote, "It was in great measure due to the indomitable courage and zeal of Mr. Field inspiring us that we went on and on until we got another cable across the gulf."8

They also contracted again with the firm of Küper and Company, but this time both to make and lay the cable, at its own risk. Field had learned his lesson and wanted no more yachting parties. The cable was successfully laid the next summer, 1856, using the steamer *Propontis*, with Samuel Canning once more the engineer in charge of the operation. That year also saw the completion, finally, of the landline across Newfoundland and the 140-mile line across the far more settled and more gentle landscape of Cape Breton Island.

The New York, Newfoundland, and London Telegraph Company had a line running from New York to Newfoundland and was thus one-third of its way, geographically, to its goal. Technologically, of course, it had barely begun, for this was the easy part. Financially, too, only a beginning had been made. While the new line would generate some revenue with traffic between St. John's and points south, that would, at best, barely cover the maintenance expenses. St. John's was a small port of no importance in and of itself.

The company had also nearly exhausted its capital. The fiasco of the first cable-laying attempt had cost the firm $351,000, all of it a dead loss. Field had personally invested $200,000 and Cooper, Taylor, and Roberts each not much less.* With their original investments nearly all sunk, the directors decided to raise additional capital with an issue of bonds, which they took up themselves. Obviously, a great deal more money was going to be needed before the mighty Atlantic could be spanned. Such sums could be raised only in the richest nation in the world. Cyrus Field sailed once more for England.

---

*To give some idea of what $200,000 meant in the 1850s, consider that $1,000 per year was enough income for a family to live a modest, middle-class life, and there were not two dozen men in all of New York City—which then had a population of over 700,000—who had a net worth of $1,000,000.

V

Raising More Capital

THE LONDON MONEY MARKET was, as yet, untapped. "Our little company," Field said proudly ten years later, "raised and expended over a quarter of a million pounds sterling before an Englishman paid a single pound."[1] Field now intended to change that and moved to form a British company.

The concept of the corporation (known as a company in Britain) is central to the financing of so large an undertaking as the Atlantic cable. Partnerships had long been the primary means of pooling capital in Western economies, but partnerships had severe disadvantages. As the need for capital increased, and more partners joined, they tended to become unwieldy and hard to manage.

The corporation came into being in the Renaissance. Indeed, except for the nation-state itself, the corporation is the most important organizational invention of the Renaissance, for without it, the modern world could not have come into being. Also developed at this time was another invention crucial to the corporation: double-entry bookkeeping. Large corporations would have been impossible to manage without it, and their profits would have been very hard to determine. The importance of good accounting was recognized very early. Ferdinand and Isabella saw to it that accountants accompanied Columbus on his voyage in 1492, to ensure they got their share of the hoped-for booty.

What makes a corporation different from a partnership is the concept of limited liability. Unlike a partnership, where every partner's entire net worth is at risk, in a joint-stock company only the amount invested could be lost, for technically a corporation is a legal "person." A corporation can be sued, it can own property, it can go broke, it can even be criminally indicted. And, of course, a corporation can be taxed. But the stockholders are *not* the corporation. They can be held to account for their own actions, but not for the actions of the corporation they own an interest in.

Thus, using this form of organization, many investors could join together to seek the potentially huge profits in exploration and distant trade without having to fear being wiped out by the equally huge risks. But whereas anyone can form a partnership with another person, only the government had the power to create a legal person in the first place. That is why the early history of corporations is so full of politics, and it would remain so until well into the nineteenth century.

In England, the Moscow Company and the East India Company were chartered by the crown in the late sixteenth century and evolved into vast, and vastly profitable, enterprises. Indeed, the East India Company would be the sovereign power in British India until the middle of the nineteenth century. Holland's East India Company would make the Netherlands the richest country in Europe in the seventeenth century.

And without the joint-stock company, the history of that most modern nation, the United States, would have taken a very different turn indeed. For one thing, many of the early colonies were founded by joint-stock enterprises seeking profits in the New World. Like so many start-up companies in a brand-new field, most were financial failures. But the Dutch West India Company, which founded New York, did very well, at least initially. It cost the Dutch West India Company twenty thousand guilders to establish New Amsterdam. In the very first year of the colony's formal existence, the company sent back to Holland more than 7,500 furs, valued at forty-five thousand guilders.

Even those colonies that had more on their agenda than profits—such as freedom to practice their religious beliefs—were expected to pay dividends. Both the Massachusetts Bay Colony and the Plymouth Colony were organized as joint-stock companies. The investors in the Plymouth Colony were not pleased, to put it mildly, when the *Mayflower* returned to England in the spring of 1621 without a cargo of goods they could sell.

The right to operate with limited liability would remain a gift of the sovereign or his representatives for centuries. But when the American colonies achieved their independence, the crown's power to establish corporations devolved on the successors to the British crown, the various state legislatures. This introduced a new element into corporation forming: the give-and-take, now-you-see-it-now-you-don't politics that characterizes legislatures.

Beginning in the 1820s, legislatures began giving up the power to form corporations. The reason was not altruistic, for it is not in the nature of human beings to give up power voluntarily. Rather, it was sheer necessity. In the entire colonial period, a grand total of only seven businesses were incorporated in the American colonies. But, as the Industrial Revolution rolled on and capital needs increased dramatically, the demand for enterprises organized as corporations increased as well. In the last four years of the eighteenth century, there were 335 businesses incorporated in the new United States. Between 1800 and 1860, the state of Pennsylvania alone incorporated more than 2,000. The state legislatures were simply unable to process all the applications for corporate charters.

As early as 1811, New York State became the first state to pass a general incorporation law, although it was restricted to companies seeking to manufacture particular items, such as anchors and linen goods. Most of the incorporations in the early years, however, involved companies that proposed to build and operate railroads, canals, and telegraph lines.

In Britain as well the Industrial Revolution caused the formation of many new corporations, either by charter from the crown or by statute. Because of the South Sea Bubble, a stock fraud in the early eighteenth century that ruined thousands, there was long-standing hostility to the concept of limited liability except for public utilities, such as railroads. But the pressure for a means of raising large sums from many people who risked no more than their investment proved irresistible. In 1844 a general incorporation law passed Parliament, and the number of companies began to increase dramatically, as did business on the London Stock Exchange.

BESIDES NEEDING THE FRESH financial pastures to be found in the United Kingdom, Field also realized that the British government had an important interest in the success of the Atlantic cable in particular and in long-distance submarine cables in general. By the mid–nineteenth century, the sun had long since ceased to set on the British Empire. But governing that empire from London was a slow process: two weeks to Canada, two months to India, three months or more to New Zealand. During the Crimean War, which was just ending, the British and French had cooperated successfully on laying a cable across the Black Sea, from Varna, in what is now Bulgaria, to Balaklava on the Crimean Peninsula, allowing the two governments to manage matters much more closely than would otherwise have been possible.* It was at the time by far the longest submarine cable in the world. But once the politicians possessed the ability to personally direct day-to-day actions in a faraway war, they naturally yearned to have instant communications everywhere, the better to gather more power into their own hands.

Back in England, Field wrote to George Villiers, Lord Clarendon, the foreign secretary, asking for a meeting.† One was quickly arranged, and Samuel Morse, whose prestige by this time was considerable, was also present to lend support. The conversation lasted an hour and Lord Clarendon showed great interest and asked many questions. Perhaps a bit startled not only by the magnitude of the enterprise but also by Field's confidence that it could be successfully completed, he asked, "But suppose you *don't* succeed? Suppose you make the attempt and fail—your cable is lost at sea—then what will you do?"

"Charge it to profit and loss," said Field in his usual direct and very American way, "and go to work to lay another."[2]

Clarendon asked Field to put his proposal in writing and intimated that the British government, which often dealt with such requests by offering "all aid short of actual help,"[3] would look with favor on this project. Field shortly received an invitation to

---

*It was ordered and laid with such haste that it was not armored at all, but was merely a copper wire insulated with gutta-percha. Nevertheless, it worked for a year, which was long enough.

†Clarendon was a professional diplomat for fifteen years before succeeding an uncle as the fourth earl of Clarendon and entering the House of Lords and politics. He remains to this day the only professional diplomat ever to be British foreign secretary. He was also one of the best. He was again serving as foreign secretary in 1870 when he died a few weeks before the Franco-Prussian War broke out. Later, Bismarck told Clarendon's daughter that if her father had lived, "he would have prevented the war."

visit James Wilson, secretary to the treasury, and spend a few days at his country house to discuss the proposal in depth.

Afterward, Wilson wrote Field a letter containing the British government's offer: to provide ships to take any further sounding that might be needed, and to consider favorably any request for ships to help lay the cable itself. Further, it agreed to pay, once the cable was working, £14,000 a year, provided that government messages had priority over all others except those of the government of the United States if it agreed to enter into a similar arrangement. The amount of £14,000 equaled the interest (at 4 percent) on the estimated cost of laying the cable — £350,000.

Field could hardly have hoped for better terms. It meant that the final route would be surveyed at no cost to the company, that the great expense of chartering, altering, and manning major oceangoing steamers to lay the cable could be avoided, and that prospective English investors would know that, once the cable was laid, the interest on their investment would be secure. Field immediately wrote to Peter Cooper to tell him of this development. He pointed out that while both ends of the cable would be in British territory, the British government expected no more than equality with the United States. Field suggested that Cooper write to President Franklin Pierce to solicit the American government for the same terms.

Meanwhile, Field formed the Atlantic Telegraph Company, chartered in London in October 1856. To raise the money, he made a tour of the country, addressing chambers of commerce and other meetings with people of means. There he presented the facts, figures, estimates, and opinions of engineers and scientists, including Samuel Morse, who had connected in series ten 200-mile telegraph circuits running between London and Birmingham. Using these, he demonstrated that as many as two hundred signals a minute could be transmitted through a 2,000-mile-long wire.

Needless to say, there was a great deal of publicity regarding so grand a scheme as well as, inevitably, many doubts and unsolicited suggestions about how best to accomplish the task. People suggested suspending the cable from balloons, or floating it just below the surface. One person thought it should be attached to a string of buoys where ships could tie up to send and receive messages. Even Prince Albert got in on the act, suggesting that the cable be enclosed in a glass tube.

Many people thought the whole project to be flatly impossible. The company received many letters pointing out that the cable would not sink to the bottom but would float at the point where its density equaled the density of the water. In fact, since water is very nearly incompressible, its density does not increase significantly with depth. The density of the water at the bottom of the Atlantic, despite the enormous pressure, is only about 2 percent more than it is at the surface.

These ideas were not confined to the scientifically illiterate. Even Astronomer Royal Sir George Airy, who ran the Royal Greenwich Observatory, stated that "it was a mathematical impossibility to submerge the cable successfully at so great a depth, and if it were possible, no signals could be transmitted through so great a length."[4]*

But the sheer grandeur of the enterprise, along with Cyrus

---

*British Astronomers Royal have a reputation for making fools of themselves with pronouncements outside their field of expertise. In 1956, Sir Richard Wooley, on assuming the post, declared that "space travel is utter bilge." A year and nine months later, the Soviet Union launched *Sputnik* and the space age began. Thirteen years later, men landed on the moon.

Field's unfeigned enthusiasm, ensured that there would be takers. Field's simple determination also helped. When one wealthy potential investor, whose passion was raising very large beagles, declined to participate, Field visited him at his kennels, hoping still to persuade him. In the course of the visit, Field asked him if he could imagine a dog so large that it stretched from England to the United States. The man replied, "Certainly, Sir; I could *imagine* anything."

"Well," replied Field, "the project of which I've spoken and tried to interest you in is essentially just such a dog. If you pinch his tail in Liverpool, he'll bark in New York."[5]

The entire stock issue, 350 shares at £1,000 each, was subscribed within a few weeks: 110 shares were taken in London, 86 in Liverpool, 37 in Glasgow, 28 in Manchester. Among the individual investors was the novelist William Makepeace Thackeray. Two American bankers long resident in London, George Peabody and C. M. Lampson, took £10,000 and £2,000 respectively.* Field himself took 88 shares, expecting to be able to sell them to investors in the United States. He was to be greatly disappointed in this and was able to dispose of only 27 of them, some of them at a loss. Thus Cyrus Field had to carry the remaining 61 shares himself, more than twice the investment of all other American shareholders combined. This investment, $305,000, together with the $200,000 he had already put into the scheme, represented the great bulk of his assets. The fortunes of both

---

*George Peabody was, along with Peter Cooper, the founder of the tradition of vast philanthropy among the American rich. He founded the Peabody Institute in Baltimore, the Peabody Museums at Yale and Harvard, and the Peabody Education Fund to help the cause of reconstruction after the Civil War by improving Southern schools. Offered a baronetcy by Queen Victoria in recognition of his generosity, he modestly declined. When he died in 1869 his banking house was taken over by his partner, Junius Spencer Morgan, and became the London branch of the House of Morgan.

Cyrus Field and the Atlantic cable had become inextricably linked. They would prosper or fail together.

WHEN CYRUS FIELD RETURNED to the United States, arriving in New York on Christmas Day, he had to go immediately to St. John's to attend to business there. The trip to Newfoundland in the dead of winter was dreadful, and Field, very susceptible to sea-sickness, soon fell ill. Regardless, once he had accomplished what he came to do (arranging for the legislature to grant to the Atlantic Telegraph Company landing privileges similar to what it had already granted to the New York, Newfoundland, and London Telegraph Company), he returned at once to New York.

There he found that he was needed immediately in Washington, where there was serious opposition to an agreement along the lines of the British one. The directors had written President Franklin Pierce, enclosing a copy of the offer of the British government and asking for similar terms. However, the British government speaks with one voice, as the prime minister and the cabinet are in firm control of both the executive departments and Parliament. It can act very quickly when it wants to. But in Washington, Congress is entirely independent of the president and has its own political considerations to take into account.

Pierce, seeing obvious advantages to the participation of the American government in the laying of an Atlantic cable, was in favor of the proposal. But he was reluctant to back it publicly, so Senator William Seward of New York (soon to be President Lincoln's secretary of state) submitted the proposal to Congress, which would have to appropriate the necessary funds. With no far-flung empire to manage and defend, many in Congress failed to see any national interest in an Atlantic cable.

Like many British intellectuals and writers (Charles Dickens, for instance), their American counterparts distrusted and disliked the effects of the Industrial Revolution in general, and such new technology as railroads, steamboats, and the telegraph in particular. Just two years earlier, the misanthropic Henry David Thoreau had written contemptuously in *Walden,* "We are in great haste to construct a magnetic telegraph from Maine to Texas; but Maine and Texas, it may be, have nothing important to communicate. We are eager to tunnel under the Atlantic and bring the Old World some weeks nearer to the New; but, perchance the first news that will leak through into the broad, flapping American ear will be that Princess Adelaide has the whooping cough."[6]

Of more concern to Congress was the strong undercurrent of Anglophobia that had run through the American psyche since the Revolution and which would not entirely disappear until the United States entered World War I on the side of Britain. This feeling was especially prevalent in the South, which had suffered grievously during the Revolution from marauding British soldiers under the likes of Bannister Tarleton and Patrick Ferguson. Senator James C. Jones of Tennessee stated forthrightly that he "didn't want anything to do with England or Englishmen."[7]

And, of course, there were some in Congress who felt that since the projectors of the Atlantic cable were all very rich men, they should pay for the privilege of having the legislation passed. As a result, Cyrus Field had to go to Washington and lobby hard to get the bill through Congress. "Those few weeks in Washington," his brother reported, "were worse than being among the icebergs off the coast of Newfoundland. The Atlantic Cable has had many a kink since, but never did it seem to be entangled in such a hopeless twist as when it got among the politicians."[8]

Many senators, of course, had no problem seeing the potential utility of the enterprise and fought hard to get the legislation passed.

Besides Senator Seward of New York, these backers included Senator Stephen A. Douglas of Illinois, whose debates with Lincoln two years hence would help delineate the coming conflict between North and South, and Senator Judah P. Benjamin of Louisiana, who would serve in the Confederate cabinet as attorney general, secretary of war, and secretary of state.*

Benjamin, addressing the Senate, confessed,

> I feel a glow of something like pride that I belong to the great human family when I see these triumphs of science, by which mind is brought into instant communication with mind across the intervening oceans, which, to our unenlightened forefathers, seemed placed there by Providence as an eternal barrier to communication between man and man. Now, Sir, we speak from minute to minute. Scarcely can a gun be fired in war on the European shore ere its echoes will reverberate among our own mountains, and be heard by every citizen in the land. All this is a triumph of science — of American genius, and I for one feel proud of it, and feel desirous of sustaining and promoting it.9

Despite such political backing, the bill cleared the Senate by a single vote and the House by not much more. It was signed into law by President Pierce on March 3, 1857, the day before he left

---

*Judah P. Benjamin's was one of the more extraordinary lives of the nineteenth century. Born a British subject in the Virgin Islands in 1811 of Jewish parents, he came to the United States at the age of two. Admitted to the Louisiana bar in 1832, he was soon one of that state's most successful lawyers. Offered a seat on the U.S. Supreme Court in 1852, he declined, and was elected a senator in 1854. After the Civil War, he escaped to England, where he was called to the English bar in 1866, quickly becoming one of its leaders and named a queen's counsel in 1872. His textbook *Law of Sale of Personal Property*, published in 1868, became the standard work on the subject for many years in both Britain and the United States. On his death in 1884, *The Times* of London wrote of him that "he carved out for himself by his own unaided exertions not one, but three [careers] of great and well-earned distinction."

office, to be replaced by James Buchanan. The act granted the New York, Newfoundland, and London Telegraph Company essentially the same deal the London government had given the Atlantic Telegraph Company in terms of subsidy, surveying, and ships.

Indeed, one promise had already been kept, as Lieutenant O. H. Berryman, U.S.N., commanding the steamer USS *Arctic*, had the previous year carefully surveyed the route along a great circle running from St. John's to the coast of Ireland.* In the spring of 1857 the Royal Navy sent Lieutenant Commander Joseph Dayman, commanding HMS *Cyclops*, to make a similar survey.

A precise knowledge of the depths to be encountered was necessary to design the cable. The reason was not that depth affects how well a submarine cable operates but that depth determines how much strain will be put on it as it is being laid. The greater the depth, the longer the length of cable that would necessarily have to hang from the ship, its weight pulling on the cable. The cable, obviously, would have to withstand the greatest strain to be encountered, so the depth determined the required strength of the cable.

The two surveys took soundings every two or three miles near shore, where the bottom was found to slope gently on the Newfoundland side and more abruptly near Ireland. Farther out, the soundings were spaced every twenty or thirty miles. The reason they were done so infrequently is that at ocean depth, the paying out and reeling in of the heavy hemp line was a time-

---

*A great circle is any circle on the surface of a sphere whose center is the same as the center of the sphere. The equator and all lines of longitude are thus great circles. Great circles are important in navigation because the shortest distance between any two points on the surface of a sphere is the arc of a great circle.

The Telegraph Plateau in the Atlantic Ocean.

consuming business, each sounding requiring almost six hours. The new soundings fully confirmed the earlier American findings of a telegraph plateau (as Matthew Maury had dubbed it) running between the two landing sites, the average depth slowly increasing between Newfoundland and Ireland.

The only serious exception to this relatively level ground was about 200 miles off Ireland, where the depth plunged in about a dozen miles from 550 fathoms to 1,750 fathoms. Although 600 feet in a mile is a steep gradient (the U.S. Interstate Highway System does not exceed 211 feet in a mile), it is no worse than occurs in many city streets, such as in San Francisco, and is not much worse than Murray Hill in New York or Holborn Hill in London.

The surveys were also able to determine the nature of the bottom as well as its depths, thanks to an ingenious mechanical device invented by Lieutenant J. M. Brooke of the United States Navy. The traditional sounding line had consisted merely of a lead weight attached to a line marked at intervals to indicate how much had been payed out. Sometimes the bottom of the lead was coated in wax, which would take an impression of the bottom. Lieutenant Brooke's "sounder" featured an iron weight much like a cannonball with a hole through its center. This was fitted over a hollow tube with a trapdoor and attached to the line by a pair of links. Allowed to fall freely through the water, when it hit bottom, the weight drove the tube into the ocean floor, grabbing a sample of it. The impact caused the links to release the weight, which was left behind, allowing the line to be hauled up quickly without fear of its breaking.*

The soundings taken by the two navies with this device revealed that the floor of the ocean for virtually the entire route was covered in ooze, consisting largely of the exoskeletons of plankton that had drifted down from the surface after death. The presence of this ooze—the dust of the sea—indicated that the bottom along the proposed route was free of scouring currents.

With the route determined, the companies formed, the capital raised, and substantial government help secured, it only remained to settle the exact design of the cable itself and order its manufacture. Indeed, by the time the American government had agreed to help, the cable was already well on its way to being completed.

---

*Brooke's invention became the standard sounding device for generations, and the ocean floor today is littered with uncounted thousands of weights that have been left behind.

VI

THE FIRST CABLE

CYRUS FIELD AND THE OTHER DIRECTORS, both in England and in the United States, were anxious to lay the cable that summer and that meant decisions had to be made quickly. This rush to get the job done was understandable. The sooner the cable was laid and working, the sooner the companies would begin to receive the $140,000 a year in payments from the British and American governments and the sooner they would begin receiving revenues from other customers.

Moreover, public interest in the enterprise was intense on both sides of the Atlantic. Newspapers and magazines not only followed actual developments closely but speculated endlessly regarding how the future would be affected by knitting the whole world together with instant communication. The *New York Herald* called the Atlantic cable "the grandest work which has ever been attempted by the genius and enterprise of man." The New York *Evening Post* thought that it would "make the great heart of humanity beat with a single pulse." An editor in Tennessee, getting quite carried away, thought that the various nations would "assemble around a common council board and engage together in diplomatic deliberations. . . . Wars are to cease. The kingdom of peace will be set up."[1]

But the rush to get the job done would prove the greatest mistake of the entire enterprise. The finest minds in electrical engineering and theory at the time were by no means in agreement as to the nature of the best design. The best way to determine the best design was experimentation, but the time pressure did not allow for much. As a result, the directors, most of whom had no competence whatever in the infant field of electrical engineering, made decisions based mostly on considerations of speed and cost. It was Cyrus Field's greatest failure as the entrepreneur of the Atlantic cable.

Samuel Morse had shown that messages could be transmitted rapidly over two thousand miles of telegraph line. To Morse and the directors, this was proof that the project was entirely feasible. But a telegraph line strung through air—which is not a conductor of electricity—is by no means a model for one strung through seawater. The fact that seawater is an excellent conductor of electricity changes everything.

It is commonly thought that electricity flows through a wire at the speed of light, about 186,000 miles a second. But that is the speed of light (or any other form of electromagnetic energy) propagating through a vacuum. Through a medium, such as air or water, the speed of light is reduced. Electricity flows through copper wire much more slowly still, often only 10 or even 1 percent of the speed of light.* The exact speed depends on the wire's capacity. The larger the wire, the more electricity is needed to "fill" it, in a way not at all dissimilar to a hose, which must fill with water before any will come out the far end.

In telegraphy's earliest days, this was no problem, because the capacity of the wires was so small that they operated at full speed nearly instantly. But when a telegraph wire is laid in sea-

---

*Of course, this is of little more than theoretical interest. A telegraphic message moving at just 1 percent of the speed of light would still cross the Atlantic in only slightly more than a second.

water, its capacity may be as much as twenty times greater, causing the signal to slow down very considerably, a problem called
retardation. There were two schools of thought about how best
to deal with the problem. One theory held that the wire should
be as small as possible. No less an authority than Michael
Faraday stated flatly that "the larger the wire, the more electricity was required to charge it; and the greater was the retardation
of that electric impulse which should be occupied in sending
that charge forward."[2]

Samuel Morse was also of this opinion, as was Dr. Edward
Whitehouse, who would soon be hired as electrician for the
Atlantic Telegraph Company. Whitehouse was a medical doctor
by profession but had done much experimenting in telegraphy. In
this dawn of the electrical age, his qualifications were equal to the
job he had. Unfortunately, he was also obstinate and always
utterly certain of the correctness of his own opinions.

Two other men who became closely associated with the
Atlantic cable at this time were of a different opinion. Charles
Bright, introduced to Cyrus Field by John Brett of the Magnetic
Telegraph Company, for which Bright worked, was appointed
chief engineer of the project, although he was only twenty-four.
The son of a successful chemical manufacturer, Bright had been
involved with telegraphy since he was fifteen. At nineteen he had
laid a complete system of telegraph wires under the streets of the
city of Manchester in a single night. By the time he was twenty
he held no fewer than twenty-four patents, some of them so
basic—such as the porcelain insulator—that they are still in use.

Bright had been interested in the possibility of an Atlantic
cable as early as 1853, and in 1855 he surveyed the west coast of
Ireland to find a suitable landing site for such a cable. He chose
Valentia Bay in the extreme southwest of Ireland, in County
Kerry, because it was sheltered from the fierce Atlantic rollers

Charles Bright

and featured a smoothly shelving, sandy shore. His choice would in the future become the site of all cables originating in Ireland.

The other person to become associated with the project, when he was invited to join the board of directors of the Atlantic Telegraph Company, was a professor at Glasgow University named William Thomson. Thomson, like his contemporaries Michael Faraday and James Clerk Maxwell, was one of the true giants in the history of science. The son of a professor of mathematics at Glasgow, he was born in 1824 and entered Glasgow University at the age of eleven. He took his degree at Cambridge, however, where he made a brilliant record. He then studied in Paris under the great French chemist and physicist Henri Victor Regnault. In 1846, at the age of twenty-two, he was appointed to the chair in natural philosophy at Glasgow University, a post he would hold for fifty-three years. At Glasgow he set up the world's

William Thomson (later Lord Kelvin)

first physics laboratory where students could do practical experiments instead of merely reading theory. It has been the basic method of teaching physics ever since.

Thomson's paper on the dynamical theory of heat, presented to the Royal Society of Edinburgh in 1851, contained the first statement of the Second Law of Thermodynamics, one of the fundamental laws by which we understand the universe.* But though William Thomson was a great theoretical physicist, who published more than three hundred papers in his lifetime, he was much more besides. As Sir Arthur Clarke put it, "If one took half

---

*C. P. Snow (1905–80), the writer and philosopher of science, regarded the Second Law of Thermodynamics as so fundamental that he divided the intellectual world into two halves: those who understood the law and its consequences and those who did not.

the talents of Einstein, and half the talents of Edison, and succeeded in fusing such incompatible gifts into a single person, the
result would be rather like William Thomson."[3]

Thomson's endlessly restless scientific imagination produced
a nearly endless stream of inventions, and in 1866 he formed the
firm of Kelvin and White to manufacture the more commercially important ones. So successful was the firm that at his death in
1907 he left an estate of £162,000, enough at that time to make
him a very rich man. In 1892, William Thomson became the first
British scientist to be raised to the peerage, when Queen
Victoria created him Lord Kelvin of Largs. He has been known
ever since as Lord Kelvin. In 1908, the year after he died, the
Kelvin temperature scale, devised by him in the 1850s, was named
in his honor.

Thomson's interest in what are known as transient electrical
currents got him interested in telegraphy. To put it simply,
Thomson wondered exactly what happened in the very brief
interval between the time an electric battery was applied to a wire
and when the current settled down to a steady value. One could
hardly ask for a more perfect example of how a talent for asking
the right question is the most precious attribute a scientist can
have. Thomson's experiments in this area would lead directly to
wireless telegraphy, then radio, and then television, a fundamental technological revolution that became one of the defining
attributes of the twentieth century.

But in the 1850s, Thomson's great question led him to experiment with how long it took a signal to go from one end of a telegraph line to another when the line was immersed in seawater. He
quickly deduced the so-called law of squares, arguing that the
amount of retardation of the signal was inversely proportional to
the square of the cable's length. In other words, a signal would

pass through a cable of a hundred miles only $\frac{1}{100}$ as fast as through the same cable ten miles long.

This information, of course, had profound implications for the Atlantic cable, which would have to be almost seven times as long as any submarine cable yet laid. Thomson set about experimenting to learn ways to lessen the impact of his law of squares. He advocated using particularly pure copper for the core of the cable and using a large-diameter core to reduce the resistance. He also set about developing instruments that would be sensitive to the least variations in electrical currents—very weak signals, in other words.

Charles Bright agreed with Thomson and advocated a cable that would have a core weighing 392 pounds per nautical mile with the same weight of gutta-percha insulation. But Edward Whitehouse and Samuel Morse disagreed with Thomson, and by the time Bright and Thomson joined the Atlantic Telegraph Company, it was too late for them to change the design of the cable. The directors were anxious to order the manufacture of the cable and, as Charles Bright's son noted in his history of the project, published in 1903, "they were perhaps over quick to recognize the difference in the capital required."[4]

The cable as it was ordered consisted of seven strands of copper wire, each .028 inches in diameter, twisted together to make a wire .083 inches in diameter. This core was then surrounded by

A sketch of the 1857 cable shown in actual size.

three separate layers of gutta-percha so that a flaw in any one layer of the insulation would be covered by another. The gutta-percha was then covered with hemp saturated with a mixture of tar, pitch, linseed oil, and wax, and that layer was covered in turn by an armoring layer of eighteen iron wires, each of seven strands, the whole cable being then coated by another mixture of tar. The finished cable, though far larger than any previously laid submarine cable, was much less substantial than what Bright and Thomson thought was needed. The core weighed only 107 pounds per nautical mile, while Bright had wanted 392 pounds per nautical mile. When Thomson tested sections of the cable, he was horrified to find that the copper used was so impure that some parts of the cable conducted electricity twice as well as other parts.

The finished cable was about five-eighths of an inch in diameter, only about as big around as a man's index finger. In air it weighed one ton per mile, while its weight in water was only 1,340 pounds per mile. It was calculated that it was strong enough to bear a weight of 6,500 pounds before breaking, equivalent to nearly five miles of its weight in water. As the water over the proposed route was nowhere greater than two and a half miles, this was thought an adequate safety margin.

The core and its insulation were made by the Gutta Percha Company, but because of the need for speed, the work of armoring the cable was divided between two companies. While the contracts with the two firms specified precise designs for the iron-wire armoring, unfortunately it failed to specify which way the iron wires should spiral around the core. As luck would have it, one firm's armoring spiraled to the left and the other's to the right. This would cause endless problems when the two halves of the cable had to be spliced together.

By July, the cable was ready. At 2,500 nautical miles in length, it consisted of 340,500 miles of copper and iron wire — more than enough to reach the moon—and 300 tons of gutta-percha. The core had cost £40 per mile, the armor £50. Altogether, the cable had cost £225,000.

In 1857 THERE WAS NOT a ship in the world capable of carrying 2,500 miles of submarine cable weighing 2,500 tons. Therefore there was no choice but to employ two ships and splice the cable together at some point. The engineers of the project, such as Charles Bright, engineer in chief, argued that it would be better to splice the cable at midocean and then have the two ships lay the cable simultaneously, one heading for Ireland, the other for Newfoundland. The advantages of this plan were twofold. First, the expedition could rendezvous at the splice point and wait if necessary for opportune weather to make the splice. Second, with both ships laying cable simultaneously, the total time needed to lay the entire cable would be halved, and this would also cut in half the chance of encountering foul weather while the operation was in progress.

The electricians, notably Samuel Morse, preferred to begin laying the cable in Ireland and making the splice when the cable was half laid. The advantage of this arrangement was that the ship could be in continuous contact with the shore during the entire operation, making sure that the signal was adequate. The directors opted for this plan.

The two ships assigned to the task of laying the Atlantic cable were the USS *Niagara* and the HMS *Agamemnon*. The *Niagara* had been laid down in 1845 to plans by the great American naval architect George Steers. In 1851, Steers would design the yacht *America* for a syndicate of wealthy New Yorkers. Its smashing victory over the entire Royal Yacht squadron that year would give the United States a dominance in international yacht racing that lasted 132 years and would give its name to the America's Cup, the oldest sporting trophy in the world.*

The *Niagara*, at 5,200 tons, was not only the largest ship in the United States Navy but perhaps the largest warship in the world at the time. She is also a perfect example of naval architecture in transition from sail to steam. Flush-decked, she had been designed from the keel up to be a steamship but was also equipped with three masts and a full suit of sails. Thanks to her powerful engine and graceful, yachtlike lines, she was easily capable of making twelve knots. And while wooden hulled, she was iron ribbed. Originally designed to carry ninety-one conventional cannons, the *Niagara* had been stripped of armament to be refitted with only twelve new, immensely powerful Dahlgren guns, invented by the American naval officer John Adolphus Dahlgren. Each was huge. Weighing fourteen tons, they required gun crews numbering twenty-five and fired explosive shells that weighed 130 pounds apiece. When she was assigned to help in the cable project, the *Niagara*'s refitting was delayed to make room for the cable.

The British entry, HMS *Agamemnon*, was a much more old-fashioned vessel. A wooden ship of the line with a displacement of 3,500 tons, the *Agamemnon*, although built later, would not have looked a bit out of place at the Battle of Trafalgar fifty-two years earlier. Only the smokestack of her steam engine, incongruously piercing her deck, would have been unfamiliar to Nelson.

---

*By no means the least of what the vast wealth creation of the nineteenth century brought about was the beginning of organized and then professional sports. Today, sports are a multibillion-dollar worldwide industry and a major means up the socioeconomic ladder for people of talent, much as the church and the military were in the preindustrial era.

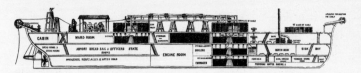

The arrangement and apparatus for stowing and paying the cable on board the Niagara.

Like most wooden battleships, she was a lubberly sailor, and would have been no match whatever for a Dahlgren-equipped *Niagara* in either sailing qualities or firepower had they ever met in battle.

But, of course, they met only in friendship. Indeed the entire 1857 attempt to lay the Atlantic cable was nearly as much an exercise in diplomacy between the two countries directly involved as it was a technological and commercial undertaking. When the *Niagara* arrived in the Thames estuary off Gravesend on May 14, the first major American warship in many years to do so, she was mobbed by visitors. Among them was Lady Franklin, widow of the British Arctic explorer Sir John Franklin. She had come to thank the U.S. Navy for the two expeditions it had mounted to search for her husband, whose last expedition had vanished after it set out in 1845 in search of the Northwest Passage.*

When the *Agamemnon* appeared and sighted the *Niagara*, the captain ordered the crew to man the yards in salute and give three cheers, which were returned in kind. The *Niagara* had been assigned to take on the segment of the cable to be loaded at

---

*Although twelve years had passed, his fate was still unknown, and it would be another two years before yet another search expedition discovered that he had perished with all his men in 1848.

Greenwich, up the Thames from Gravesend, while the *Agamemnon* would load the half that was waiting at Birkenhead, across the River Mersey from Liverpool. But the *Niagara* proved too large for the dock at Greenwich, so the two ships switched assignments and the *Niagara* made for Birkenhead, stopping first at Portsmouth to have still more of her bulkheads removed to accommodate the cable. There she was also fitted with an iron cage over the screw to prevent the cable from becoming entangled in it if the ship had to back for any reason. The cage was promptly dubbed a "crinoline," after the lady's undergarments that were made of iron bands and used to hold out the hoop skirts that had exploded into fashion the previous year.

The *Niagara* arrived at Birkenhead on June 22 and was again nearly mobbed by curious visitors while the labor of loading the cable began. It required thirty men at a time, working for three weeks, to accomplish this, as the cable had to be coiled layer upon layer (these were called "flakes") with great care so as not to kink or become tangled. But while the sailors labored, the officers that could be spared were feted again and again in Liverpool. The mayor gave a dinner, as did the Chamber of Commerce. On the Fourth of July the city's American community, which was con-siderable given the volume of Anglo-American trade—largely American cotton and British manufactures—entertained them.

The *Agamemnon* was simultaneously taking on her half of the cable at Greenwich and, when the job was completed, her officers and crew, along with many others interested in the undertaking, were entertained at an outdoor party at the house of Sir Culling Eardley, Bart., the tents and food provided by Glass, Elliot & Co., which had manufactured that half of the cable.*

As always, there were numerous speeches and toasts (*The Times* of London noted that the sailors soon exhausted their allot-ted three pints of beer drinking the toasts). Besides Sir Culling, Samuel Morse also spoke, and Cyrus Field read a letter from President James Buchanan inviting Queen Victoria to send the first message through the cable should it prove successful.

There was keen anticipation among the press and the public in Britain regarding the cable, made all the keener because the first

---

*The Times of London reported on the party extensively, including the careful arrangement of the seating to reflect Victorian notions of class. "By an admirable arrangement," it noted, "the guests were accommodated at a vast semi-circular table, which ran round the whole pavilion, while the sailors and workmen sat at a number of long tables arranged at right angles with the chord, so that the gen-eral effect was that all dined together, while at the same time sufficient distinction was preserved to satisfy the most fastidious."

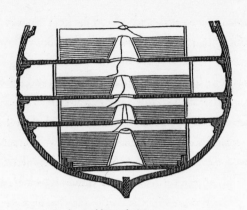

Storage of coils of cable on the Niagara.

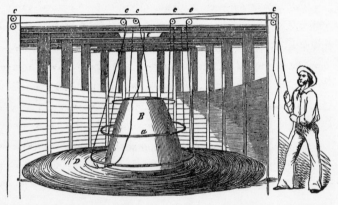

Large iron rings (A) prevent the cable from kinking as it is drawn up over the cone (B) by pulleys (C) and through the hatchway (E) until the remaining coil of cable (D) had been payed out.

reports of the great Indian Mutiny, which had begun on May 10 that year, were only now beginning to arrive, a full two months later. A cable could have brought the news in minutes.

In North America as well, there was deep interest in the completion of what was universally regarded as a wonder in an age of wonders. In Albany, New York, Edward Everett, former Harvard professor, congressman, governor of Massachusetts, secretary of state, and U.S. senator, and one of the great orators of his age, spoke of it at a dedication ceremony.

> I hold in my hand a portion of the identical electric cable, given me by my friend Mr. [George] Peabody, which is now in progress of manufacture to connect America with Europe. Does it seem all but incredible to you that intelligence should travel for two thousand miles, along those slender copper wires, far down in the all-but-fathomless Atlantic, never before penetrated by aught pertaining to humanity, save when some foundering vessel has plunged with her helpless company to the eternal silence and darkness of the abyss? Does it seem, I say, all but a miracle of art, that the thoughts of living men—the thoughts that we think up here on the earth's surface, in the cheerful light of day—about the markets and the exchanges, and the seasons, and the elections, and the treaties, and the wars, and all the fond nothings of daily life, should clothe themselves with elemental sparks, and shoot with fiery speed, in a moment, in the twinkling of an eye, from hemisphere to hemisphere, far down among the uncouth monsters that wallow in the nether seas, along the wreck-paved floor, through the oozy dungeons of the rayless deep . . . ?5*

---

*The nineteenth century was one of the golden ages of oratory, but because the speeches could not be recorded in other than written form, the orators, like the actors of that era, are largely forgotten today unless remembered for other reasons, such as Daniel Webster. But Edward Everett, the greatest American orator of his time, achieved a sort of perverse immortality with his speechmaking. It was his odd fate to deliver the main address at the dedication of the National Cemetery at Gettysburg, Pennsylvania, on November 19, 1863. His now utterly forgotten two-hour speech immediately preceded—and served as the perfect foil for—the two-minute speech of President Abraham Lincoln that has since been read and memorized by uncountable millions.

WHEN THE *NIAGARA* AND THE *AGAMEMNON* were loaded with 1,250 miles of cable each, they rendezvoused at Queenstown (now Cobh) on the southern coast of Ireland, then (and still) that country's leading transatlantic port. There, the two main ships anchored about a third of a mile apart, and one end of the cable on the *Niagara* was ferried across to the *Agamemnon*. The two halves were spliced temporarily together. A telegraph transmitter was attached to one end and a galvanometer—an instrument for measuring electric current—to the other. Doubtless to everyone's relief, the current flowed perfectly through all 2,500 miles of the Atlantic cable.

Besides the two major ships *Niagara* and *Agamemnon*, the fleet consisted of their escorts, USS *Susquehanna* and HMS *Leopard*, both side-wheelers. There were also three smaller steamers, HMS *Advice*, HMS *Willing Mind*, and HMS *Cyclops*, under Captain Dayman, who had surveyed the route the previous spring and would be in charge of any more soundings that might be required.

As the armada approached Valentia Bay, crowds of people could be seen standing on the hills to witness the most exciting thing to come to County Kerry in decades. Valentia Bay, though ideally suited as the European anchorage for the Atlantic cable, is located in one of the most remote corners of the British Isles. County Kerry was still largely Gaelic speaking in the 1850s. It also has some of the most majestic scenery in all Europe, where peninsulas that rise to Ireland's highest mountains run far out into the sea before dropping off sometimes in sheer cliffs for hundreds of feet to meet the great Atlantic rollers that crash at their feet.

Once anchored, the cable fleet was greeted by the lord lieutenant of Ireland, George Howard, Earl of Carlisle, who had come down from Dublin, and by Sir Peter Fitzgerald, Knight of Kerry, the major local landowner.*

The first task was to bring the end of the cable ashore. The fleet arrived too late on the evening of August 4 to begin work, and the next morning there was too much wind, but finally the operation got under way about two in the afternoon. The shore ends of the cable—ten miles on the Irish side and fifteen on the Newfoundland side—were much more heavily armored than the oceanic part of the cable, to protect them from waves, rocks, currents, and errant anchors. As a result they weighed fully nine tons per mile, and wrestling the end out of the *Niagara*'s hold, onto a launch, and into shore was no easy task.

The diplomatic aspect of the entire operation can be easily seen in the fact that it had been deliberately arranged for American sailors to haul the cable ashore in Ireland, where it would be officially presented to the lord lieutenant as the queen's representative. On the other side of the Atlantic, British sailors brought the cable ashore.

The lord lieutenant stood on the beach with his staff and other dignitaries while Valentia Bay, a reporter wrote, "was studded with innumerable small craft, decked with the gayest

---

*The earl of Carlisle was of the great Howard family, among whom are numbered the dukes of Norfolk and the earls of Suffolk and Effingham. A poet and political reformer, he was deeply interested in helping what today would be called juvenile delinquents. He built a model reformatory on the grounds of one of his estates. His principal residence was Castle Howard, in Yorkshire, one of Britain's most famous stately homes, best known today, perhaps, as the setting for many scenes in *Brideshead Revisited*.

Sir Peter Fitzgerald was of an Anglo-Irish family, one even more ancient than the Howards and which dominated southwestern Ireland for centuries. In the middle of the thirteenth century, one of his ancestors dubbed three of his sons hereditary knights, the only hereditary knighthoods in British history. Sir Peter was very popular with his tenants, for, unlike many Irish landlords in the nineteenth century, he lived on his estate on Valentia Island and built substantial homesteads for many of these tenants.

bunting—small boats flitted hither and thither, their occupants cheering enthusiastically as the work successfully progressed. . . . [A]nd when at length the American sailors jumped through the surge with the hawser to which it was attached, his Excellency was among the first to lay hold of it and pull it lustily to the shore."[6]

A minister said a prayer and then the lord lieutenant spoke of his hopes for what the cable might accomplish. "I believe the great work now so happily begun will accomplish many great and noble purposes of trade, of national policy, and of empire." But he also noted both a more human dimension to the cable's potential and the great calamity that had befallen Ireland in the previous decade when the potato crop had failed. Two million people—a quarter of the population—had died in those terrible years of starvation or disease, or been forced to emigrate to escape them. The earl continued,

> You are aware—you must know, some of you, from your own experience—that many of your dear friends and near relatives have left their native land to receive hospitable shelter in America. . . . If you wished to communicate some piece of intelligence straightaway to your relatives across the wide world of waters—if you wished to tell those whom you know it would interest in their heart of hearts, of a birth, or a marriage, or, alas, a death, among you, the little cord, which we have hauled up to the shore, will impart that tidings quicker than the flash of lightning. Let us indeed hope, let us pray that the hopes of those who have set foot on this great design, may be rewarded by its entire success; and let us hope, further, that this Atlantic Cable will, in all future time, serve as an emblem of that strong cord of love which I trust will always unite the British islands to the great continent of America.[7]

When Carlisle had finished, he said that the usual three cheers were entirely insufficient for this occasion and asked for at least a dozen, which echoed off the green, surrounding hills.

Cyrus Field then took his turn and spoke briefly. "I have no words," he confessed, "to express the feelings which fill my heart tonight—it beats with love and affection for every man, woman, and child who hears me. I may say, however, that, if ever at the other side of the waters now before us, any one of you shall present himself at my door and say that he took hand or part, even by an approving smile, in our work here today, he shall have a true American welcome. I cannot bind myself to more, and shall merely say: 'What God has joined together, let no man put asunder.' "[8]

Not surprisingly, perhaps, Cyrus Field found that he couldn't sleep in his cabin on the *Niagara* that night, and went up to pace the deck beneath a cloudless sky and a myriad of stars. Running the length of the deck was the paying-out machinery designed to control the speed at which the cable moved from the cable tier in the front of the ship to the large wheel at the stern over which it passed before it slipped into the sea. Men were to be stationed along its length to see that the cable ran smoothly and not too fast, and braking mechanisms were at the ready to slow it down if necessary.

At dawn the next morning, the fleet of three British (*Agamemnon, Leopard,* and *Cyclops*) and two American ships (*Niagara* and *Susquehanna*) got under way, the cable trailing out behind the USS *Niagara*. But before the fleet had proceeded five miles, the heavy shore cable caught in the machinery and parted. There was nothing to do but return to Valentia Harbor and start over. Boats from the *Willing Mind* ran a hawser under the cable to lift it to the surface, and the *Willing Mind* then proceeded seaward, lifting the

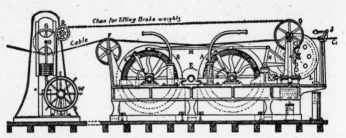

Bright's 1858 paying-out gear.

cable as she went until the broken end was reached. This was then spliced to the cable on the *Niagara* and the expedition began again, moving at no more than two knots.

When the end of the shore cable was reached, it was spliced onto the ocean cable and the splice was lowered by hawser to the ocean floor, while buoys marked the point. The *Niagara* slowly increased its speed as it began paying out the lighter ocean cable, and the paying-out machinery groaned and rumbled with a "dull, heavy sound." But the sound was a reassuring one, telling everyone that the project was proceeding as planned. "If one should drop to sleep," Matthew Field noted, "and wake up at night, he has only to hear the sound of 'the old coffee-mill,' and his fears are relieved, and goes to sleep again."[9] The cable was allowed to pay out a little faster than the ship was moving so that it could follow the contours of the sea bottom dropping away toward the ocean abyss.

The weather on Saturday, August 8, was perfect and spirits were high. Communication between the *Niagara* and Valentia was nearly continuous, with Dr. Whitehouse onshore and Samuel Morse, Cyrus Field, and William Thomson on board, although Morse was soon incapacitated by seasickness. By midnight, Saturday, eighty-five miles of cable had been laid and the depths of the water was a little over two hundred fathoms. The next morning the switch from the cable coiled on the aft deck to the one coiled forward was made without incident and the ship's speed was increased to five knots.

On Monday morning, beginning about 4:00 A.M., the ship began passing over the steep incline where the sea bottom plunged from 500 to 1,750 fathoms within eight miles. The paying-out machinery continued to function, although the cable more than once was thrown off the wheels, owing to the fact that the groove in the wheels where the cable was supposed to run was badly designed and not large enough. Tar from the cable also tended to build up in these grooves, reducing the room still further.

When this happened, the ship had to be halted and the cable clamped until it could be fitted back into place. As Charles Bright, the engineer in charge, reported, if nothing else, at least this laying to in deep water without paying out more cable, while tons of it hung from the stern proved that such an operation was possible. The cable, of course, has been designed to withstand the strain of such a weight. Still, it was a relief to determine that it could actually do so.

As a longer and longer length of cable hung from the back of the *Niagara*, it was necessary to apply the brakes more firmly to prevent the cable from paying out too fast. Charles Bright noted

The principle of the brake used on the cable.

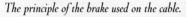

that while seven hundred pounds of braking pressure had been sufficient at first to keep the speed of the cable at the right rate, at this point it became necessary to increase the braking pressure to fifteen hundred pounds while the ship and the cable ran at five and five and a half knots respectively.

At noon on Monday, the *Niagara* was 214 miles from shore and had payed out 255 miles of cable. The breeze began to freshen and a swell to get up. The ocean floor was now some two thousand fathoms below and the braking pressure had been increased to two thousand pounds.

Then, at nine o'clock that evening, the cable suddenly ceased to function. There had been an ongoing stream of messages between the ship and the shore since the flotilla had departed Valentia Bay, but now there was nothing. Samuel Morse, miserably seasick, was nonetheless roused from his bed, but he was unable to make the cable come alive again, despite two and a half hours of effort. All seemed lost and the engineers were preparing to cut the cable and attempt to bring it back on board until the flaw could be found, an operation for which there was little precedent.

Suddenly the cable came back to life and began to function in a perfectly normal fashion. No one, then or since, has a satisfactory explanation for what happened. Morse himself suggested that in the course of the cable's jumping out of its wheeled track and being placed back in it, it had been strained and that had opened a breach in the gutta-percha insulation, allowing seawater to short out the circuit. Then when the cable reached the bottom, the seam had reclosed, either because of how it lay or because the water pressure forced the seam closed.

Regardless, the cable was once again working, and as John Mullahy, of the *New York Tribune*, one of several reporters on board the *Niagara*, wrote, "The glad news was soon circulated throughout the ship, and all felt as if they had new life."[10]

That evening, the speed of the cable slowly began to gain on the speed of the ship, and by nine o'clock Monday night, it was still running out at the rate of five and half knots while the ship had slowed to three knots. Bright ordered the strain to be increased to twenty-five hundred pounds. It was not enough. As Bright stated in his official report to the directors of the company, with "the wind and sea increasing, and a current at the same time carrying the cable at an angle from the direct line of the ship's course, it was found insufficient to check the cable, which was at midnight making two and a half knots above the speed of the ship, and sometimes imperiling the safe uncoiling in the hold."[11] He increased the braking force to three thousand pounds at two A.M., and then to thirty-five hundred pounds.

Once the cable had mysteriously come back to life, Cyrus Field went to bed in his cabin. But at a quarter to four, just as dawn was beginning to lighten the eastern horizon, he heard a sudden rush of voices from on deck, including the ominous "Stop her! Back her!"[12]

Within seconds, Charles Bright was at his door. "The cable's gone," he told him.[13]

Like many disasters, this one had come about through simple bad luck.

Charles Bright reported,

I had up to this attended personally to the regulation of the brakes, but finding that all was going well, and on it being necessary that I should be temporarily away from the machine—to ascertain the rate of the ship, to see how the cable was coming out of the hold, and also to visit the electrician's room—the machine was for the moment left in charge of a mechanic who had been engaged from the first in its construction and fitting, and was acquainted with its operation.

In proceeding towards the fore part of the ship I heard the machine stop. I immediately called out to relieve the brakes, but when I reached the spot the cable was broken. On examining the machine—which was otherwise in perfect working order—I found that the brakes had *not* been released, and to this, or to the hand-wheel of the brake being turned the wrong way, may be attributed the stoppage and consequent fracture of the cable.[14]

The stern of the *Niagara* had been in the trough of a wave. As the ship rose, the strain on the cable increased and the brake should have been eased to compensate, as Bright had been doing all day. The sudden combination of increased strain and continued braking was too much, and the cable parted with a sound like a pistol shot. The end of the cable vanished in an instant and plunged to the depths of the sea, nearly two and a half miles down.

Field wasted no time on regrets. He ordered the other ships to be signaled to heave to and for their captains to come on board the *Niagara*. Captain Hudson, commanding the *Niagara*, wrote that the disaster "made all hands of us through the day like a household or family which had lost their dearest friend, for officers and men had been deeply interested in the success of the enterprise."[15] Someone lowered the *Niagara's* flag to half-mast and the other ships quickly followed suit.

Field asked that the *Agamemnon, Niagara,* and *Susquehanna* remain at the site for a few days "to try certain experiments which will be of great value to us."[16] These included splicing the two halves of the cable together while at sea and laying it from both ships simultaneously. When the experiments or, perhaps more accurately, the practice runs were completed, the fleet was to proceed to Plymouth. He asked the *Cyclops's* captain to take precise soundings, which he did and then left for Valentia Bay. Meanwhile, Field boarded HMS *Leopard* and made for Portsmouth immediately, where he intended to call a special meeting of the board of directors of the Atlantic Telegraph Company.

The directors quickly decided that it would not be possible to make a second attempt that year. Four hundred miles of cable had been irretrievably lost and there was not enough time left to manufacture more and remount the expedition before the gales of autumn made the attempt too hazardous. When the *Niagara* and the *Agamemnon* arrived in Plymouth, they were ordered to unload the remaining cable and return to duty in their respective navies.

No one directly connected with the enterprise was discouraged by this failure, regrettable as it was. As with any new technology, there was much to learn and many failures to be expected as engineers worked their way up the learning curve. Already three things were clear. First, the paying-out machinery, which had been hastily designed and built without taking into account the far greater weights involved in the Atlantic cable, needed to be redesigned and safety mechanisms employed to make it more difficult for human error to come into play. Second, the various maneuvers involving the ships and the cable needed to be rehearsed thoroughly. And third, the crew needed to be better trained in the techniques of cable laying. The chief engineer could not monitor it all by himself.

But while those directly involved were sanguine about the eventual outcome, public opinion began to shift due to the accident. The I-told-you-sos, of course, were having a field day and a parody of "Pop Goes the Weasel!" began making the rounds:

Pay it out! Oh, pay it out
As long as you are able:
For if you put the damned brake on,
Pop goes the cable!

~VIII~

"AND LAY THE ATLANTIC CABLE IN A HEAP"

IT WAS FORTUNATE that the Atlantic Telegraph Company had raised the £100,000 in additional capital needed among its own stockholders in Britain, for there would be no capital available in the United States. As the *Niagara* and *Agamemnon* were attempting to lay the cable, a financial hurricane swept through both New York and the American economy. When Cyrus Field returned to New York in mid-September, it was blowing more fiercely than ever.

The American economy, which had endured a severe depression from 1837 to 1843, had begun to expand again in the mid-1840s. This expansion was greatly helped by the demands of the Mexican War, which erupted in 1846. A rapidly growing and increasingly assertive United States sought to reach the Pacific Ocean under the doctrine of "manifest destiny." When it was unable to negotiate the purchase of Mexico's vast, nearly empty northern provinces, both countries declared war. After a brilliant campaign to seize the Mexican capital led by Zachary Taylor, who would be elected president in 1848 because of it, Mexico had no choice but to seek terms. In 1848, in exchange for $15 million and the assumption of American claims against Mexico, the latter ceded what is today roughly the southwest quarter of the continental United States.

Even before the peace treaty had been ratified, however, gold was discovered east of San Francisco. As gold miners rushed in from all over the world, a flood of gold, then legal tender throughout the world, was added to the American economy. In 1847 the country had produced only 43,000 ounces of gold, mostly as a by-product of other mining. The next year, however, gold production reached 484,000 ounces and the year after that, as miners flooded into California by the tens of thousands, output was 1,935,000 ounces.

This flood of new wealth set off one of the country's great economic expansions. Production of pig iron, a good measure of the economy as the Industrial Revolution developed, increased from 63,000 tons in 1850 to 883,000 tons only six years later. Coal production more than doubled; railroad trackage tripled. Nearly as many corporations were formed in the 1850s as had been formed in the previous half century.

The exuberant boom had made possible the initial funding of a project so fraught with uncertainty as the Atlantic cable. The sky was the limit in 1854. By 1857 the situation had changed.

The American economy at this time lacked a fundamental institution of a modern economy, a central bank, one of the primary functions of which is to provide a sound monetary system and to guard the national economy against excesses of the business cycle by disciplining other banks as necessary.

President Andrew Jackson, however, a Jeffersonian Democrat when it came to economic policy, hated all banks. He had killed the Second Bank of the United States in 1836 by vetoing a renewal of its charter. There would not be a successor institution until 1913, when the Federal Reserve was created. With state banks subject only to the erratic control of state governments, they tended to make too many loans in good times and issue too much paper money. In bad times they tended to fail in large numbers.

Thus economic booms in the United States in the nineteenth century tended to run out of control, producing more goods and services than could be absorbed by the economy. This overproduction inevitably ended in catastrophic crashes, followed by deep and long-lasting depressions.

By August 1857, after nine years of rapid expansion, the signs of economic trouble were everywhere. Six thousand textile looms in New England were idle, and shipping crowded the country's harbors for lack of cargo. On August 24, when a New York bank suddenly closed its doors in insolvency, prices on Wall Street began to decline sharply. By September the weaker banks and brokerages were collapsing. And when it was learned that the steamer *Central America*, on its way from Panama with four hundred passengers and, of more concern to Wall Street, $1.6 million in California gold, had sunk with fearful loss of life off Cape Hatteras, Wall Street crashed.

Despite the fact that communication to Europe still required a minimum of ten days, the panic soon spread to Europe, where the Banks of France and England jacked up interest rates to protect their currencies, and European investors began pulling out of American securities. American banks were forced to suspend payments in gold. The American economy would be mired in depression for another four years.

Field arrived home just in time to learn that his own paper firm, Field and Company, was in extremis, with $600,000 in debts and no way to pay them because many of its accounts receivable were temporarily uncollectible. Field immediately called his creditors together and offered them notes paying 7 percent interest. The reputation for straight shooting that Field had acquired when he paid off the debts of Root and Company although he was not legally obligated to do so now stood him in good stead. His creditors accepted his proposal readily.

And it must have come as a slightly cheering fact that when he arrived at his offices, he found them full of men who were not there to collect money but desperate to utilize the services of the line to St. John's, Newfoundland, in hopes of getting word to or from London a few days quicker. This powerfully confirmed Field's prediction that there was a strong market for instant transatlantic communication.

His personal affairs back in order, at least for the moment, Field went to Washington and asked the secretary of the navy for two things, the use of the *Niagara* and *Susquehanna* in the next attempt, and to have the engineering officer of the *Niagara*, William E. Everett, released from his naval duties so that he could work full-time on the project. Everett, who had helped to design the engines of the *Niagara*, had watched the original paying-out

William Everett

machinery being installed on the ship and had taken professional notice of how inadequate it was. He found it too complex, and with a braking system that applied pressure too abruptly. Everett had very clear ideas on how to improve matters and Field wanted his undivided assistance.

Fortunately, the government was cooperative and Secretary Isaac Touchy was able to hand Field an official letter, saying, "There, I have given you all you asked."[1] A week after getting Touchy's response, Field and Everett boarded the Cunard liner *Persia,* built in 1855 and the largest and fastest transatlantic steamer of the time, capable of crossing the Atlantic in a little over nine days.

In London, William Everett was named chief engineer of the company and proceeded to the engineering firm of Eston and Amos in Southwark, across the Thames from London proper, where he spent six months designing and supervising the construction of new paying-out machinery. Its most innovative feature was a new brake system based on a design by J. G. Appold, a wealthy amateur mechanic.* The new brakes could be set with a maximum braking pressure according to circumstances and the brakes would automatically release if the strain exceeded that limit, preventing the cable from breaking and being lost. Demonstrated to Field, to Charles Bright, the chief engineer of the project, and to a group of outside engineers, including Isambard Kingdom Brunel, the greatest of British engineers who was later to play a major part in making the Atlantic cable successful, the new brake "seemed to have the intelligence of a human being, to know when to hold on, and when to let go."[2]

The entire apparatus weighed only one-fourth what the pre-

---

*In mid–Victorian times the gentlemen tinkerer, nearly extinct today, was still a common phenomenon.

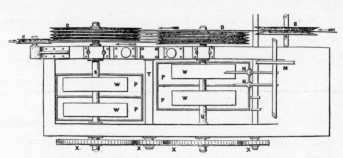

Everett's paying-out machine.

vious machinery did and took up only one-third the deck space.
How well William Everett did his job can be measured by the
fact that the paying-out machinery of cable-laying ships ever since
has followed his brilliant design.

William Thomson had also been striving to improve matters
for the next attempt, along with maintaining his full workload at
the University of Glasgow. It was financially impossible to scrap
the entire cable and build a better one designed to his specifications, as too much had already been invested in it. But he carefully measured the electrical conductivity of the new cable and
rejected any portions that were not up to the highest level.

Thomson had also been thinking long and hard about the
speed with which messages could be transmitted across a very long
submarine cable. The electric impulses that make up the dots and
dashes of Morse code tend to smear out as they travel—the dots,
in effect, turning into dashes and the dashes into longer dashes.
This made it necessary to transmit fairly slowly. The solution of
Dr. Whitehouse to this problem was to force as strong a pulse of
electricity as possible down the wire. Dr. Whitehouse was still the
company's head electrician, although Cyrus Field was listening to
him less and less and was working to have him removed from his
position. He had already bought out Whitehouse's financial interest in the company for $8,000.

Whitehouse, whose personality brooked no dissent, bitterly resented being edged out and railed at "the frantic fooleries of the Americans in the person of Mr. Cyrus W. Field."

Thomson tried the opposite approach, a more sensitive device to receive the signal, capable of reacting even to very faint ones. One day Thomson happened to be twirling his monocle around its chain and he observed how it sent a spot of light dancing around the walls of the room. In a classic "eureka moment," he had the answer: the mirror galvanometer, a device that has been fundamental to submarine telegraphy ever since.

A galvanometer, first devised in 1802, is an instrument for detecting electric current. A simple galvanometer consists of a needle that is deflected in the presence of a magnetic field induced by current, much as a compass needle is deflected by the earth's magnetic field. The stronger the current, the greater the deflection.

But in a long submarine cable, immersed in a conducting medium—saltwater—the current is often very low, sometimes no more than ten microamperes. (The current in a standard incandescent lightbulb is about 100,000 times as great.) The standard galvanometers then available were often inadequate to detect a signal coming through a cable that would be two thousand miles long. So Thomson—half Einstein, half Edison—developed a much better one. He took a very small magnet and attached a tiny mirror to it. Both together weighed no more than a grain.* He suspended the magnet from a silk thread and set it in the middle of a coil of very thin insulated copper wire.

When the faint current flowing through the cable was allowed to flow through the copper coil, it created a magnetic field. This caused the magnet, with its attached mirror, to deflect. Thomson simply directed a beam of light from a shaded lamp onto the mir-

*There are 437.5 grains to an ounce.

ror and allowed its reflection to hit a graduated scale. As the varying current in the copper coil caused the magnet and mirror to deflect, the variations were greatly magnified by the reflected beam of light and, using Morse code, could be read easily by an observer at rates up to twenty words per minute, ten times the old rate.

It was universally acknowledged as a brilliant piece of applied physics. James Clerk Maxwell, the greatest theoretical physicist of the age, even wrote a poem about it:

> The lamplight falls on blackened walls,
> And streams through narrow perforations;
> The long beam trails o'er pasteboard scales,
> With slow, decaying oscillations.
> Flow, current, flow! Set the quick light-spot flying!
> Flow, current, answer, light-spot! Flashing, quivering, dying.[3]*

WHEN FIELD HAD FIRST RETURNED to England, the Atlantic Telegraph Company had offered him the post of general manager of the company at the handsome salary of £1,000 a year. Field declined the salary but accepted the post.

The directors obviously did not share Dr. Whitehouse's opinion of Field. "The directors cannot close their observations to the shareholders," they wrote in a report, "without bearing their warm and cordial testimony to the untiring zeal, talent, and energy that

---

*Maxwell's poem, which is several stanzas long, is a parody of one of the lyrics found in Tennyson's *The Princess: A Medley*, first published in finished form in 1853. Tennyson, an enormously popular poet in an age when poetry was still widely read, had been named Poet Laureate in 1850 on the death of Wordsworth. Thirty years later, *The Princess* would serve as the source of Gilbert and Sullivan's *Princess Ida*, which Gilbert, typically, called "a respectful operatic perversion" of Tennyson's poem.

have been displayed on behalf of this enterprise by Mr. Cyrus W. Field, of New York, to whom mainly belongs the honor of having practically developed the possibility and of having brought together the material means for carrying out the great idea of connecting Europe and America by a submarine telegraph."[4]

As spring approached, the second expedition to lay the Atlantic cable began to lumber into action. The *Niagara* and the *Agamemnon* arrived in Plymouth and began to take on the cable that had been off-loaded the previous autumn and to take on as well the new cable that had been manufactured. The cable tiers of the two ships, large round spaces designed to hold the endless coils, had been fitted with cast-iron cones to make stowing the cable easier and safer for the cable. The process took all of April and half of May to accomplish, as only about thirty miles of cable could be stowed per day in each ship, forty-man crews working around the clock.

In late May, Field got word that the crew of the *Susquehanna* had been struck with yellow fever in the West Indies, and the ship was quarantined. Field immediately called on Sir John Pakington, first lord of the admiralty,* and asked for help. Although the Royal Navy was so pressed for ships itself that it was chartering some, Pakington promised to see what he could do and within a few hours Field learned that HMS *Valorous*, under Captain W. C. Oldham, had been dispatched to take the place of the *Susquehanna*.

After the cable had been loaded, the fleet proceeded in late May to deep water in the Bay of Biscay. There it practiced the various techniques of cable laying that the previous year's attempt had shown needed work: splicing the two halves of the cable together while at sea, laying simultaneously from both ships, and switching

*The equivalent position in the American government would be secretary of the navy.

from one coil to another. Because the two halves of the cable had been armored with iron wires having different lays—one spiraled to the right, the other to the left—an ordinary splice was not sufficient because the armoring would tend to untwist. So an elaborate wooden and iron coupling device was created to ensure that the splice would hold together. It was no small affair, being some twelve feet long and weighing three hundred pounds.

This time Cyrus Field had decided to follow the advice of his engineers, rather than his electricians, and lay both halves simultaneously. On Thursday, June 10, back from its successful practice run-through, the fleet of four vessels—USS *Niagara*, HMS *Agamemnon*, HMS *Gorgon*, and HMS *Valorous*—set off from Plymouth for a point in the mid-Atlantic, 52°2′ N, 33°18′ W, from which the cable laying would begin. William Everett was on board the *Niagara* and Charles Bright and Samuel Canning on board *Agamemnon*, as was William Thomson, substituting again for Edward Whitehouse, who had, once again, pled illness. Samuel Morse was absent, having preferred to remain in New York rather than face the certain seasickness to which he was so subject. His place was taken by C. V. de Sauty, who had helped lay the cable across the Black Sea during the Crimean War.

June has a reputation, not always deserved, for having the Atlantic's finest weather. And as the fleet departed it could hardly have been better. "Never did a voyage begin with better omens," Henry Field reported. "The day was one of the mildest of June, and the sea so still, that one could scarcely perceive, by the motion of the ship, when they passed beyond the breakwater off Plymouth harbor into the channel, or into the open sea. At night, it was almost a dead calm. The second day was like the first. There was scarcely wind enough to swell the sails."[5]

The third day, too, was fair and mild. But on Sunday, June 13, the wind picked up and the barometer began to fall. The cable fleet was sailing into the heart of one of the worst storms in the history of the north Atlantic. *Gorgon* and *Valorous*, small, nimble, and well founded, could expect to deal with even the worst of weather. But *Niagara* and *Agamemnon* were each burdened with 1,500 tons of cable, tremendous loads for ships of only 5,200 and 3,500 tons respectively.

Indeed, the *Agamemnon* had had to store 250 tons of the cable on its forward deck, high above the ship's center of gravity. This not only brought the ship down by the head but reduced her stability as well. And the cable stored below had diminished the size of the ship's coal bunkers, so 160 tons of coal had been stored in sacks on the main and gun decks, toward the stern to help balance the weight of the cable forward. These great and concentrated masses made sailing the *Agamemnon*, not a good sailor in the best of circumstances, extremely tricky.

As the storm strengthened, the ships shortened sail and spread out to give each other sea room. *Gorgon* and *Valorous* were soon lost to view amid the deepening gloom of "a wretched mist—half rain, half vapor"[6] and the increasingly towering waves. But *Niagara* and *Agamemnon* were to sight one another occasionally throughout the ordeal.

On Saturday, June 12, the *Agamemnon* had sailed under a cloud of canvas. By Sunday evening, however, she was under only close-reefed top sails and foresails. As the wind blew the foam off the waves, the sea soon resembled "one vast snow drift, the whitish glare from which—reflected from the dark clouds that almost rested on the sea—had a tremendous and unnatural effect, as if the ordinary laws of nature had been reversed."[7]

As the wind howled through the rigging and water hissed beneath her keel, the *Agamemnon* added the great variety of noises made by a large wooden ship in a heavy sea. As Nicholas Woods, a reporter from *The Times* of London, noted, "The massive beams under her upper deck coil cracked and snapped with a noise resembling that of small artillery."[8] As the ship worked, the massive weight of the coil of cable on her deck held the seams of the planking open and allowed water to pour into the ship. Sailors were soon set to the backbreaking labor of manning the pumps, which added their mournful, steady clanking to the cacophony of sounds.

"The sea," Woods continued, "kept striking with dull, heavy violence against the vessel's bows, forcing its way through the hawse-holes and ill-closed ports with a heavy slush; and thence, hissing and winding aft . . . in some five or six inches of dirty bilge," which soaked everything, until "the ship was almost as wet inside as out."

"Such was Sunday night," the reporter noted, "and such was a fair average of all nights throughout the week, varying only from bad to worse."

The next morning, although the barometer was lower still and the wind and seas higher, the sun suddenly appeared and shone brilliantly for half an hour. But "during that brief time [the wind] blew as it had not often blown before. So fierce was this gust that it drowned out every other sound, and it was almost impossible to give the watch the necessary orders for taking in the close-reefed foresail, which, when furled, almost left the *Agamemnon* under bare poles, though still surging through the water at speed."

On Tuesday the barometer rose somewhat, to 29.3 inches, and there was enough sun to allow an officer to take a sextant reading and determine that the ship was 563 miles from the mid-Atlantic rendezvous. But during the afternoon, "the *Agamemnon* took to violent pitching, plunging steadily into the trough of the sea as if she meant to break her back and lay the Atlantic cable in a heap."

The *Agamemnon* was now rolling steadily thirty degrees to either side of vertical, often more. The masts, cutting arc after arc across the leaden sky, worked their wire shrouds loose, and it was possible to see the masts bending at every oscillation. Under the circumstances, there was nothing to be done, but everyone on board knew what the loss of the masts would mean. They acted as a counterweight to dampen the *Agamemnon*'s wild rolling, and if the masts went, the ship would certainly capsize. Other things were also coming loose. The ship's largest boat, stored on her main deck, had partly broken free of its restraints and swung from side to side of the ship as she rolled, threatening to crush everyone on deck as the crew wrestled the multi-ton vessel into submission.

For six days the storm raged around the cable fleet before diminishing. On Saturday, June 19, "the barometer seemed inclined to go up and the sea to go down; and for the first time that morning since the gale began, six days previous, the decks could be walked with tolerable comfort and security." It was not to last, and "during a comparative calm that afternoon the glass fell lower, while a thin line of black haze to windward seemed to grow up into the sky, until it covered the heavens with a somber darkness, and warned us that the worst was yet to come. There was much heavy rain that evening, and then the wind began, not violently, nor in gusts, but with a steadily increasing force, as if the gale was determined to do its work slowly but do it well."

On Sunday, June 20, the storm unleashed a fury such as few sailors ever see and even fewer live to tell about. The captain feared that the coil on the deck, working against its restraints, might break lose and smash through the side, undoubtedly causing the ship to founder. He ordered that the coil's restraints be reinforced.

But the 250 tons of cable on deck still sought freedom like some chained beast as the ship rolled from side to side, straining the timbers that supported her (and held the ship together) to the breaking point. One of the beams supporting the lower deck cracked under the strain and had to be shored up with jack screws by the ship's carpenters. As the storm relentlessly deepened, everyone not actively employed jammed themselves in corners or held on to beams to prevent being flung about like dice in a cup.

The reporter wrote,

> At ten o'clock, the *Agamemnon* was rolling and laboring fearfully, with the sky getting darker, and both wind and sea increasing every minute. At about half-past ten o'clock three or four gigantic waves were seen approaching the ship, coming slowly on through the mist nearer and nearer, rolling on like hills of green water, with a crown of foam that seemed to double their height. The *Agamemnon* rose heavily to the first, and then went down quickly into the deep trough of the sea, falling over as she did so, so almost to capsize completely on the port side. There was a fearful crashing as she lay over this way, for everything broke adrift, whether secured or not, and the uproar and confusion were terrific for a minute, then back she came again on the starboard beam in the same manner, only quicker and still deeper than before.

The *Agamemnon in the storm.*

The off-watch officers in the ward room rushed to the ward room door, leading to the main deck. "Here," the reporter continued, "for an instant, the scene almost defies description. Amid the loud shouts and efforts to save themselves, a confused mass of sailors, boys, and marines, with deck-buckets, ropes, ladders, and everything that could get loose, and which had fallen back again to the port side, were being hurled again in a mass across the ship to starboard. Dimly, and only for an instant, could this be seen, with groups of men clinging to the beams with all their might, with a mass of water, which had forced its way in through ports and decks, surging about."

Then, in the midst of this chaos, "with a tremendous crash, as the ship fell still deeper over, the coals stowed on the main deck broke loose, and smashing everything before them, went over among the rest to leeward."

Many of the sacks in which the coal was stored ripped open and choking coal dust blotted out what feeble light there was. "But the crashing could still be heard going on in all directions, as the lumps and sacks of coal, with stanchions, ladders, and mess-tins, went leaping about the decks, pouring down the hatchways, and crashing through the glass skylights into the engine-room below." As the *Agamemnon* lurched once more sickeningly to port, the coal stored on the gun deck gave way as well.

Captain Preedy ordered hands to bring the ship onto the other tack. It was a perilous maneuver, at best, in such conditions, but the ship would ride more easily on that course. It would also take them away from the rendezvous, but he had no choice. Hundreds of tons of coal and debris were surging back and forth belowdecks as the ship rolled from side to side at up to forty-five degrees. Not even the mighty oak sides of a ship of the line could take that for long.

The *Niagara*, far larger and better designed, fared better in the storm and would suffer only moderate damage. But those on board were powerless to help the *Agamemnon* and could only look on when she was in sight.

Preedy thought also of cutting loose the cable on deck and wrestling it over the side. From the beginning of the storm, that had been Cyrus Field's worst nightmare, and every time the *Niagara* sighted the *Agamemnon*, he looked with great anxiety to see if the cable—and thus all hope for a successful expedition— still rode on *Agamemnon*'s deck. Jettisoning the cable would have increased the speed with which the ship rolled, but lessened the amplitude. Preedy decided to wait on that fateful step, at least for now.

The officers shouted orders to the crew with precision and the crew tried to obey, but the conditions made it almost impossible. "The officers were quite inaudible," the reporter wrote, "and a wild, dangerous sea, running mountains high, heeled the great ship backward and forward so that the crew were unable to keep their feet for an instant, and in some cases were thrown clear across the decks in a fearful manner." Yet somehow the maneuver was accomplished and the *Agamemnon* steadied on her new course.

Officers went below to tend to the many injured and to get the coal and other loose objects under control. Although the cable stored on the deck had held fast, the cable in the hold had not. "The top kept working and shifting over from side to side, as the ship lurched, until some forty or fifty miles were in a hopeless state of tangle, resembling nothing so much as a cargo of live eels, and there was every prospect of the tangle spreading deeper and deeper as the bad weather continued."

Once matters were under some sort of control belowdecks, Captain Preedy decided to return to the port tack, on course for the rendezvous. But as the ship went about, she "fell into the trough of the sea again, and rolled so awfully as to break her wastesteampipe, filling her engine room with steam, and depriving her of the services of one boiler when it was sorely needed."

"The sun set," the reporter continued, woeful but undaunted, "upon as wild and wicked a night as ever taxed the courage and coolness of a sailor. . . . [O]f all on board none had ever seen a fiercer or more dangerous sea than raged throughout the night and the following morning, tossing the *Agamemnon* from side to side like a mere plaything among the waters."

The next morning it had become increasingly clear that the *Agamemnon* was nearing the end of her endurance. "The masts were rapidly getting worse, the deck coil worked more and more with each tremendous plunge, and, even if both these held, it was evident that the ship itself would soon strain to pieces if the weather continued so."

There were three possibilities available to Captain Preedy. He could return to the starboard tack, and risk broaching to (presenting the side of the ship to the waves and putting her in great danger of capsizing) in the process. Or he could attempt to run before the storm, placing the wind at his stern. This would greatly ease the motion of the ship, but it was a dangerous course. Unless very carefully handled—and perhaps even then—a wave might stave in the stern galleries and send hundreds of tons of water surging along the decks, driving the ship underwater to her doom. Finally, he could jettison the deck coil, an expedient he would adopt only to save his ship from certain destruction.

He ordered the *Agamemnon* over onto the starboard tack, and while the head came about, the ship wallowed in the trough of a wave and "all the rolls which she had ever given on the previous day seemed mere trifles compared with her performance then. . . . It really appeared as though the last hour of the stout ship had come, and to this minute it seems almost miraculous that her masts held on. Each time she fell over, her main chains went deep under water. The lower decks were flooded, and those above could hear by the fearful crashing—audible amid the hoarse roar of the storm—that the coals had gotten loose again below, and had broken into the engine room, and were carrying all before them."

Providentially, the *Agamemnon* rallied and came onto the starboard tack. The wisp of sail she was showing filled and she regained her way, but her motion remained as bad as on the old tack. Preedy then "succumbed to the storm he could neither conquer nor contend against. Full steam was got on, and with a foresail and a fore-topsail to lift her head the *Agamemnon* ran before the storm, rolling and tumbling over the huge waves at a tremendous pace."

ALL STORMS MUST END, and finally, as the *Agamemnon* galloped before the wind, this one, too, began to moderate. By 4:00 A.M. Tuesday, the weather had abated enough to allow the *Agamemnon* to resume course toward the rendezvous, now some two hundred miles farther away than it had been Sunday morning.

It was Friday, June 25, sixteen days since leaving Plymouth, when the *Agamemnon* arrived at 52°2′ N, 33°18′ W. "As we approached the place of meeting," the relieved and doubtless exhausted reporter for *The Times* of London reported, "the angry sea went down. The *Valorous* hove in sight at noon; in the afternoon the *Niagara* came in from the north; and at even the *Gorgon* from the south: and then, almost for the first time since starting, the squadron was reunited near the spot where the great work was to have commenced fifteen days previously—as tranquil in the middle of the Atlantic as if in Plymouth Sound."

Cyrus Field had a boat lowered and rowed over to the *Agamemnon*. He found the deck a shambles, the white planking on

which the Royal Navy had so long prided itself black with coal
dust. While miraculously no one had been killed, forty-five men
were in sick bay with injuries ranging from broken arms to tempo-
rary insanity, and a hundred miles of the cable was in a laocoönian
tangle. But by the next afternoon, the weather now cold and foggy
but calm, the two ships were ready to begin the cable laying.

They maneuvered stern to stern and the *Niagara*'s cable was
brought aboard the *Agamemnon*, where the two halves were
spliced. A bent sixpence was placed in the splicing device for luck
and it was lowered over the side. But hardly had the two ships
begun to move apart when, as Captain Hudson recorded in the
Niagara's log, "the cable, being hauled in the wrong direction,
through the excitement and carelessness of one of the men, caught
and parted in the *Niagara*'s machinery."9

The loss of cable was insignificant, and the two halves were
soon respliced and the laying began again. Field estimated that in
five days the two ships would reach their respective destinations and
the job would be done. But the next morning at 3:30 A.M., when
each ship had laid about forty miles of cable, Professor Thomson
came out onto the deck of the *Agamemnon* to say that the line had
gone dead. At nearly the same instant, Field and C. V. de Sauty,
the electrician on board *Niagara*, announced the same thing. It was
assumed that the cable had parted somehow, but each ship assumed
that the problem must be on the other ship, as everything was normal as far as they could see. Following Field's orders, both ships
returned to the rendezvous. When they came in sight of each
other, *Niagara* ran up the signal, "How did the cable part?"10

"This was astounding," Nicholas Woods reported. "As soon
as the boats could be lowered, Mr. Cyrus Field, with the electricians from the *Niagara*, came on board, and a comparison of logs
showed the painful and mysterious fact that, *at the same second of
time*, each vessel discovered that a total fracture had taken place at

a distance of certainly not less than ten miles from each ship: in fact as well as can be judged, at the bottom of the ocean."[11]

But while the cause was—and remains—an utter mystery, there was no time for thorough investigations. Supplies were running short. Field called the engineers and electricians—Bright, Canning, and Thomson from the *Agamemnon*, and Everett and de Sauty from the *Niagara*—to a conference on board the *Niagara*. They agreed to the following proposal, written down by Field: "Should any accident occur to part the cable before the ships have run a hundred miles from rendezvous . . . the ships shall return to rendezvous and wait eight days, when, if the other ships do not appear, then proceed to Queenstown."[12] Everyone signed the agreement.

The cable was spliced between the two ships once more—the eighty miles that had been laid were lost when each ship returned to the rendezvous—and again the two ships parted. And at first all went well. "The scene at night was beautiful," someone on the *Niagara* reported, "Scarcely a word was spoken; silence was commanded, and no conversation allowed. Nothing was heard but the strange rattling of the machine as the cable was running out. The lights about the deck and in the quarter deck circle added to the singularity of the spectacle; and those who were on board the ship describe the state of anxious suspense in which all were held as exceedingly impressive."[13]

On the *Agamemnon,* too, all went well. "At nine o'clock by ship's time" on Tuesday evening, Nicholas Woods reported, "when 146 miles had been payed out and about 112 miles' distance from the rendezvous accomplished, the last flake but one of the upper deck coil came in turn to be used."[14]

This meant that the transfer from the deck coil to the main coil would have to be effected in the middle of the night. But the maneuver had been rehearsed in the Bay of Biscay and there was every confidence that it would go without a hitch. The ship and the paying-out machinery were slowed in anticipation, the screw being reduced from thirty revolutions per minute to twenty while the paying-out machinery slowed from thirty-six to twenty-two. There was barely a ton of strain registering on the *Agamemnon*'s dynamometer when, near midnight, the cable suddenly snapped and the end vanished over the stern.

On board the *Niagara*, they knew at once that the cable had parted and that this time the break was very near the *Agamemnon*. Field waited for a while, hoping against hope that the signal would return, as it had returned the previous year. When it did not, he reluctantly ordered the *Niagara* to make for Queenstown, in accordance with the agreement, as both ships had gone more than one hundred miles from the rendezvous. Captain Preedy made a different decision, deciding that as he was only fourteen miles over the hundred-mile limit, he would seek to rendezvous once again. He headed back to the rendezvous point and searched for the *Niagara* for a week, often in thick fog, until he, too, made for Queenstown, arriving a week after the *Niagara*.

It would turn out that the lowest section of the cable in the deck coil, lying directly on the deck, had been damaged by the violent storm that had nearly cost the *Agamemnon* her life. It was this that was paying out when the rupture occurred. The storm had gotten in its final lick after all.

Cyrus Field, though certainly discouraged, was not dismayed. It was, for all the disappointment, only a technical misfortune that had robbed him of success, not some basic flaw in the scheme itself. He was determined to try again, but he needed to persuade the directors.

⋆⇒VIII⇐⋆

LIGHTNING THROUGH
DEEP WATERS

ONCE THE *AGAMEMNON* RETURNED to Queenstown, Field, Thomson, and Bright left immediately for London to meet with the board of directors of the Atlantic Telegraph Company. The news of the second failure had, of course, reached London ahead of them. The board, which had been in high hopes in the spring, was now sunk in gloom and defeat. Henry Field, who undoubtedly heard directly from his brother, reported that "the feeling—to call it by the mildest name—was one of extreme discouragement. They looked blankly in each other's faces. With some, the feeling was one almost of despair."[1]

The chairman of the board, Sir William Brown, a prominent Liverpool merchant deeply involved in transatlantic trade and founder of the great banking house of Brown, Shipley, had been unable to attend the board meeting in London, but he sent a downbeat message. "We must all deeply regret our misfortune," he wrote, "in not being able to lay the cable. I think there is nothing to be done but dispose of what is left on the best terms we can."[2] He then wanted to pay the bills, distribute what was left over to the stockholders, and liquidate the corporation.

Field protested this unbridled pessimism but found that it was seconded by the vice chairman of the board, T. H. Brooking, who was present and who had until this last setback been among the enterprise's most ardent supporters. He announced that he was "determined to take no further part in an undertaking which had proved hopeless, and to persist in which seemed mere rashness and folly."3 Brooking then walked out of the meeting and sent in his resignation the next day, as did Brown.

Field, staggered by these defections, addressed the rest of the board and called on all his considerable ability as a salesman to persuade them to carry on for one more try. He pointed out that while three hundred miles of cable had been lost in the most recent expedition, enough still remained to do the job. He also pointed out that the ships and crews were in Queenstown, the cable loaded. They needed only coal and supplies and some repairs to the battered *Agamemnon* to be ready for another attempt. Bright and Thomson backed him strongly, pointing out that valuable experience had been gained from the previous failures, making success more probable this time.

Realizing that there was little to lose, except a couple of thousand miles of submarine telegraph cable of dubious resale value, the board went along with Field and his technical staff. Field and Samuel Gurney, a board member, then went immediately to the admiralty and orders were telegraphed to the cable fleet to get ready to sail as soon as possible. Field, Thomson, and Bright returned to Queenstown, and on Saturday, July 17, the fleet sailed out of the Cove of Cork to try once more.

This time there were no cheers from shore. Indeed, as Nicholas Woods of *The Times* of London noted, "There was apparently no notice taken of their departure by those on shore or in the vessels anchored around them. Everyone seemed impressed with the conviction that we were engaged in a hopeless enterprise; and the squadron seemed rather to have slunk away on some discreditable

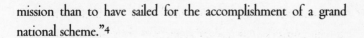

mission than to have sailed for the accomplishment of a grand national scheme."4

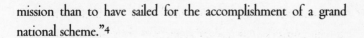

THE EXPEDITION WAS EXPERIENCING THE EFFECTS of one of the most important but least recognized developments of the Industrial Revolution, public opinion. The rotary steam press, invented in the 1820s, had made possible newspapers printing thousands, and soon tens of thousands, of copies an issue. When this capability was combined with the idea of politically independent newspapers appealing to a mass audience in the next decade, the modern media were born.

In the 1840s, the telegraph made it possible for newspapers throughout a wide area to cover the same story in real time, both powerfully molding public opinion and expressing it in such devices as letters to the editor. It was soon almost impossible to imagine a world without this new force. "The daily newspaper," wrote the *North American Review* in 1866, "is one of those things which are rooted in the necessities of modern civilization. The steam engine is not more essential to us. The newspaper is that which connects each individual with the general life of mankind."5

Many people, deeply disappointed at the latest failure and influenced by the pessimism of many members of the board, had decided that the Atlantic cable was a wild-goose chase. Even many on board the ships thought "they were going on a fool's errand; that the company was possessed by a kind of insanity, of which they would be cured by another bitter experience."6

The voyage to the rendezvous was relatively uneventful, with periods of both dead calm and stiff breeze. "Indeed," wrote Nicholas Woods, "until the rendezvous was gained, we had such a succession of beautiful sunrises, gorgeous sunsets, and tranquil moonlit nights as would have excited the most enthusiastic admiration of any one but persons situated as we were."7

The *Agamemnon* did not reach the rendezvous until Wednesday, July 28, five days after the *Niagara* got there. The next day was perfect for making the splice. Cyrus Field, not given to flights of language, recorded the details and the task ahead in his journal.

> Thursday, July 29, latitude fifty-two degrees twenty-seven minutes west, longitude thirty-two degrees twenty-seven minutes west. Telegraph fleet all in sight; sea smooth; light wind from SE to SSE; cloudy. Splice made at one P.M. Signals through the whole length of the cable on board both ships perfect. Depth of water fifteen hundred fathoms; distance to the entrance of Valentia harbor eight hundred and thirteen nautical miles, and from there to the telegraph-house the shore end of the cable is laid. Distance to the entrance of Trinity Bay, Newfoundland, eight hundred and twenty-two nautical miles, and from there to the telegraph-house at

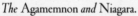

The Agamemnon *and* Niagara.

the head of the bay of Bull's Arm, sixty miles, making in all eight hundred and eighty-two nautical miles. The *Niagara* has sixty-nine miles further to run than the *Agamemnon*. The *Niagara* and *Agamemnon* have each eleven hundred nautical miles of cable on board, about the same as last year.[8]

Later he described his own mood. "When I thought of all that we had passed through, of the hopes thus far disappointed, of the friends saddened by our reverses, of the few that remained to sustain us, I felt a load at my heart almost too heavy to bear, though my confidence was firm and my determination fixed."[9]

The splice, weighted with a thirty-two-pound cannonball after its lead sinker fell off, was lowered over the stern of the *Agamemnon*, and after 210 fathoms of cable was payed out, the two ships began their journeys. The ships stayed in constant contact with an exchange of current sent every ten minutes, alternately by each ship. In addition, every time ten more miles of cable had been payed out by one ship, the information was conveyed to the other, via the cable, as were messages regarding momentary slow-downs or stoppages to allow for switching coils or splicing.

Within a few hours the ships increased their speeds, the *Agamemnon* moving at about five knots while the cable payed out at six knots. Shortly after the speed was increased, a "very large whale"[10] appeared off the *Agamemnon*'s starboard bow, apparently making directly for the cable. "Great was the relief of all," Woods wrote, "when the ponderous living mass was seen slowly to pass astern, just grazing the cable where it entered the water — but fortunately without doing any mischief."[11]

At eight o'clock that night, the first crisis occurred. A damaged portion of the cable was discovered just a mile or two from what was being payed out at that instant, and it would be only about twenty minutes until it vanished over the side. Workmen frantically tried to repair the bad spot, for everyone was most reluctant to stop either the ship or the cable, for fear of causing a break. But then the cable suddenly stopped transmitting.

The engineers immediately assumed that the damaged part must be responsible, and ordered it cut out and a splice begun. The ship was stopped and no more cable was payed out than absolutely necessary to prevent its breaking when the *Agamemnon*'s stern rose in a swell. *The Times* of London reporter wrote,

> The main hold presented an extraordinary scene. Nearly all the officers of the ship and of those connected with the expedition stood in groups about the coil, watching with intense anxiety the cable as it slowly unwound itself nearer and nearer the joint, while the workmen worked at the splice as only men could work who felt that the life and death of the expedition depended upon their rapidity. But all their speed was to no purpose, as the cable was unwinding within a hundred fathoms; and, as a last and desperate resource, the cable was stopped altogether, and for a few minutes the ship hung on only by the end. Fortunately, however, it was only for a few minutes, as the strain was continually rising above two tons and it would not hold much longer. When the splice was finished the signal was made to loose the stoppers, and it passed overboard in safety.[12]

The signal from the *Niagara,* however, did not return. The fault, after all, lay elsewhere between the two ships. Field, on board the *Niagara*, was told by de Sauty that the cable was sound but that the current was just not coming through. There was nothing to do but wait and hope. A reporter on board from the *Sydney Morning Herald* wrote, "The scene in and about the electrical room was such as I shall never forget. The two clerks on duty, watching, with the common anxiety depicted on their faces, for a propitious signal; Dr. Thompson, in a perfect fever of nervous excitement, shaking like an aspen leaf, yet in mind keen and collected, testing and waiting. . . . Mr. Bright, standing like a boy caught in a fault, his lips and cheeks smeared with tar, biting his nails and looking to the professor for advice. . . . The eyes of all were directed on the instruments, watching for the slightest quiver indicative of life."[13]

And life returned. Just as had happened the previous year, the cable suddenly began functioning again, as mysteriously as it stopped. It was later determined that the problem lay with one of the batteries aboard the *Niagara*.

The next morning another crisis hit as a sextant reading showed that the *Niagara* was paying out more cable than the distance covered would warrant. At that rate, there would not be enough cable to reach Newfoundland. This time, there was both a simple explanation and an easy solution. The tons of iron that made up the armoring of the cable was affecting the compass readings and the *Niagara* was seriously off course.* Field ordered HMS *Gorgon* to run ahead of the *Niagara* and lead the way.

On the *Agamemnon* the chief problem was fuel. The ship had used more fuel than expected on its way out to the rendezvous, owing to headwinds, and now encountered them again while traveling in the opposite direction. "During the evening," Nicholas Woods reported, "topmasts were lowered, and spars, yards, sails, and indeed everything aloft that could offer resistance to the wind, were sent down on deck. Still the ship made but little way, chiefly in consequence of heavy sea, though the enormous quantity of fuel consumed showed us that if the wind lasted, we

---

*Although the Earth's magnetic field is very large—eight thousand miles between positive and negative poles—it is very weak. An ordinary household magnet is stronger. As iron increasingly replaced wood in ship construction, means had to be devised to shield the compass from the effects of the ship's own magnetic field. These means would be developed in the 1870s by—who else?—William Thomson.

should be reduced to burning the masts, spars, and even the decks, to bring the ship into Valentia."[14]

But the next day the wind came around to southwest and allowed the ship to conserve fuel by hoisting some sail. The southwest wind, alas, promptly blew up into a full gale and Bright, Canning, and the other engineering staff had to be on constant vigil to see that the brakes were relieved at the right moments to prevent the cable from snapping. Everyone on board half expected to hear the gun announcing disaster at any moment. After all, as Nicholas Woods explained, the cable "in comparison with the ship from which it was payed out, and the gigantic waves among which it was delivered, was but a mere thread."[15] But the thread held, "only leaving a silvery phosphorescent line upon the stupendous seas."[16]

By this point the ships were well past the point of no return; there was not enough cable remaining for another try if it broke again.

The *Niagara* was having an easier time of it, and Field's journal records steady progress, with hardly anything more exciting than sighting a Cunard liner en route from Boston to Liverpool and exchanging greetings by signal flag. The *Agamemnon* also encountered a ship, but she required rather more emphatic communications. "About three o'clock on Tuesday morning [August 3]," Woods wrote, "all on board were startled from their beds by the loud booming of a gun. Everyone—without waiting for the performance of the most particular toilet—rushed on deck to ascertain the cause of the disturbance."

It was not, as many must have feared, that the cable had parted yet again. Instead, HMS *Valorous* had rounded to and was firing gun after gun across the bow of an American bark steering a course that would take her directly astern of the *Agamemnon* and across the cable. This quickly got the attention of those aboard the American vessel, which put the sails aback and hove to until the cable expedition was lost to sight.

"Whether those on board her considered that we were engaged in some filibustering expedition," Nicholas Woods wrote, "or regarded our proceedings as another outrage upon the American flag, it is impossible to say."[17]*

As the *Agamemnon's* escort was fending off the American vessel, the ship to which she was tethered by a line now some sixteen hundred miles long was approaching North America. The *Niagara* sighted icebergs later that day, some of them a hundred feet high and wreathed in "a corona of fleecy white clouds."[18] The icebergs confirmed what sextant readings indicated: they were near the Grand Banks, the great fishing grounds off Newfoundland. A sounding was taken and bottom was found only two hundred fathoms down. The *Niagara* had reached the continental shelf. Sixteen hundred miles away, at five o'clock that evening, ship's time, the *Agamemnon* reached the steep rise from the telegraph plateau, and at ten o'clock she also found bottom on the continental shelf.

With a far shorter length of cable hanging from the sterns of the two ships, and thus much less weight, the danger of its suddenly snapping was greatly reduced, and hope began to grow quickly on each ship that success was, at last, at hand.

Nicholas Woods wrote,

By daylight on the morning of Thursday, the 5th, the bold rocky mountains which entirely surround the wild and picturesque neighborhood of Valentia rose right before us at a few miles distance. Never, probably, was the sight of land more welcome, as it brought to a successful termi-

---

*The War of 1812, which had largely been caused by the refusal of the Royal Navy to respect the American flag at sea, was still well within living memory in the late 1850s.

nation one of the greatest, but at the same time most difficult, schemes
which was ever undertaken. Had it been the dullest and most melancholy swamp on the face of the earth that lay before us, we should have
found it a pleasant prospect; but as the sun rose behind the estuary of
Dingle Bay, tinging with a deep, soft purple the lofty summits of the
steep mountains which surround its shores, illuminating the masses of
morning vapor which hung upon them, it was a scene which might vie
in beauty with anything that could be produced by the most florid imagination of an artist.[19]

A day earlier, on Wednesday, August 4, the *Niagara* made
landfall at Newfoundland. "Made land off entrance to Trinity
Bay at eight AM," wrote Cyrus Field in his journal. "Entered
Trinity Bay at half-past twelve. . . . At five PM saw Her Majesty's
steamer *Porcupine* coming to us."[20]

All on board the *Niagara* were quite as moved by the sun setting behind Newfoundland as those on the *Agamemnon* were by
the sunrise over Ireland. "All who have visited Trinity Bay,"
wrote James Mullahy of the *New York Herald*, ". . . with one
consent allow it to be one of the most beautiful sheets of water
they ever set eyes upon. Its color is very peculiar—an inexpress-
ible mingling of the pure blue ocean with the deep evergreen
woodlands and the serene blue sky. Its extreme length is about
eighty miles, its breadth about thirty miles, opening boldly into
the Atlantic on the northern side of the island. At its southwestern shore it branches into the Bay of Bull's Arm, which is a
quiet, safe, and beautiful harbor, about two miles in breadth and
nine or ten in length, running in a direction north-west."[21]

Captain Otter of the *Porcupine* came aboard the *Niagara* to act
as pilot as the ship made her way up Trinity Bay, and at 1:45 A.M.,
Thursday, August 5, the ships anchored at the head of Bull's Arm
Bay. Although it was the middle of the night, Field immediately

had a boat lowered and had himself rowed ashore, where he went looking for the telegraph house about a mile inland. More or less lost in the dark, he finally found it and walked into the pitch dark house, where he could hear the quiet breathing of men sleeping.

He shouted for them to wake up, yelling, "The cable is laid!"[22] A light was lit and the men stared at him in astonishment. They were maintenance men and no operator able to send a message was present. The nearest point on the line where that was possible was fifteen miles away and Field asked for a volunteer to go there, taking messages. The first was to his wife, Mary. "Arrived here yesterday. All well. The Atlantic telegraph cable successfully laid. Please telegraph me here immediately."[23] Next he sent messages to his father and to the Associated Press and Peter Cooper, giving brief details and, characteristically, sharing the credit widely among the captain of the *Niagara*, the electricians and engineers; "and, in fact, every man on board the telegraph fleet has exerted himself to the utmost to make the expedition successful. By the blessing of Divine Providence it has succeeded."[24] Finally he telegraphed President Buchanan, telling him of the success, that a message from Queen Victoria would soon be sent and afterward the line would be held open for his reply.

"At a quarter-past five AM," Field wrote in his journal, with remarkable lack of emotion, "telegraph cable landed. At six, end of cable carried into the telegraph-house, and received very strong currents of electricity through the whole cable from the other side of the Atlantic. Captain Hudson, of the *Niagara*, then read prayers and made some remarks. At one PM, her Majesty's steamer *Gorgon* fired a royal salute of twenty-one guns."[25]

Remote as Valentia Bay is from the rest of Ireland, it was metropolitan compared with the nearly uninhabited wilderness of Trinity Bay, Newfoundland. And when HMS *Valorous* steamed ahead and fired a gun to alert the population to the arrival of the telegraph fleet, "there was a general desertion of the place, and hundreds of boats crowded around us," Nicolas Woods reported. The knight of Kerry was absent in Dingle, across Dingle Bay from Valentia, but he soon arrived in a Royal Navy gunboat. The end of the cable was loaded into boats from the *Valorous* and brought ashore.

"The end was seized by the jolly tars," wrote Woods, "and run off with." But they were not to have the honor of bringing to land the Atlantic cable uncontested, and a good-humored scuffle took place for possession of the cable, between men from the ship and men from on shore, including the knight of Kerry, who waded into the surf and promptly fell. Finally the cable was brought ashore by Charles Bright, Samuel Canning, and the drenched knight of Kerry, and laid in a prepared trench.

Even before the cable was attached to the landline awaiting it, Charles Bright telegraphed a message to the board of directors in London, a message instantly passed on to the press. "The *Agamemnon* has arrived in Valentia, and we are about to land the cable. The *Niagara* is in Trinity Bay, Newfoundland. There are good signals between the ships."

A WORLD-TRANSFORMING EVENT had taken place and, thanks to that event, the entire Western world, from the Missouri River in North America to the Volga in Russia, learned of it nearly simultaneously. The reaction was titanic.

George Templeton Strong, the New York diarist and Field's neighbor on Gramercy Park, was much given to viewing the glass as half empty. But he wrote in his diary on August 5 that "all Wall Street stirred up into excitement this morning, in spite of the sultry weather, by the screeching newsboys with their extras.

The *Niagara* has arrived at Trinity Bay with her end of the telegraph cable in perfect working order. But the *Agamemnon* has not yet linked her end at Valentia. . . . The transmission of a single message from shore to shore will be memorable in the world's history for though I dare say this cable will give out before long, it will be the first successful experiment in binding the two continents together, and the communication will soon be permanently established."[26]

Five days later he reported, "Everybody all agog about the Atlantic Cable. Telegraph offices in Wall Street decorated with flags of all nations and sundry fancy pennons besides, suspended across the street. Newspapers full of the theme, and of demonstrations the event has produced from New Orleans to Portland. . . . Newspapers vie with each other in gas and grandiloquence. Yesterday's *Herald* said that the cable (or perhaps Cyrus W. Field, uncertain which) is undoubtedly the Angel in the Book of Revelation with one foot on sea and one foot on land, proclaiming that Time is no longer. Moderate people merely say that this is the greatest human achievement in history."[27]

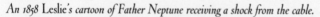

An 1858 Leslie's *cartoon of Father Neptune receiving a shock from the cable.*

Field's brother David wired, "Your family is all at Stockbridge and well. The joyful news arrived here Thursday, and almost overwhelmed your wife. Father rejoiced like a boy. Mother was wild with delight. Brothers, sisters, all were overjoyed. Bells were rung, guns fired; children let out of school, shouted, 'The cable is laid! The cable is laid!' The village was in a tumult of joy. My dear brother, I congratulate you. God Bless you."[28]

In London, *The Times,* not a newspaper easily carried away or noted for a pro-American stance, wrote that "since the discovery of Columbus, nothing has been done in any degree comparable to the vast enlargement which has thus been given to the sphere of human activity. . . . The Atlantic is dried up, and we become in reality as well as in wish one country. . . . The Atlantic Telegraph has half undone the declaration of 1776, and has gone far to make us once again, in spite of ourselves, one people."[29]

When Queen Victoria heard the news she forthwith conferred a knighthood on Charles Bright, only twenty-six, and said that she would have knighted Cyrus Field as well had he been a British subject. Because the queen was then abroad, the dubbing ceremony was performed by the earl of Carlisle as lord lieutenant of Ireland. President Buchanan telegraphed Field in Trinity Bay to congratulate him "with all my heart upon the success of the great enterprise with which your name is so honorably connected."[30] Besides the president, Field was inundated with messages from governors, bishops, the governor general of Canada, and even one from an old acquaintance whom he had met long before on his travels in western New York State selling paper for his brother Matthew: "Beech Tree, chief of Oneida tribe, honors the white man whom the great spirit appoints to transmit his lightning through deep waters."[31]

And, of course, the success had to be marked with banquets, balls, parades, oratory, and much bad poetry. In St. John's, Field was feted with all of this, while Bright, Thomson, and the others were treated equally in Dublin and then London.

The great American preacher Henry Ward Beecher astutely noted the fact that the Atlantic cable was an achievement of the age in which he lived and could have been accomplished in no other. "We are gathered to express our joy," he told his congregation, "at the apparent consummation of one of those enterprises which are peculiar, I had almost said to our generation — certainly to the century in which we live. Do you reflect that there are men among you tonight, men here, who lived and were not very young before there was a steamboat on our waters? Ever since I can remember steamboats have always been at hand. There are men here who lived before they beat the water with their wheels. And since my day railroads have been invented."[32]

The public, naturally, was wild for news and messages to begin moving across the cable, but Field, as early as August 7, tried to dampened expectations. "We have landed here in the woods," he wired the Associated Press, "and until the telegraph instruments are perfectly adjusted, no communications can pass between the two continents; but the electric currents are received freely. You shall have the earliest intimation when all is ready, but it may be some days before everything is perfected. The first through message between Europe and America will be from the Queen of Great Britain to the President of the United States, and the second his reply."[33]

Finally, on August 16, Queen Victoria's message was received.

To the President of the United States, Washington:

The Queen desires to congratulate the President upon the successful completion of this great international work, in which the Queen has taken the deepest interest.

Fireworks at City Hall in honor of the Atlantic telegraph.

The Queen is convinced that the President will join with her in fervently hoping that the electric cable will prove an additional link between the nations, whose friendship is founded upon their common interest and reciprocal esteem.

The Queen has much pleasure in thus communicating with the President, and renewing to him her wishes for the prosperity of the United States.

The President replied in kind.

When the news of the messages was released, still more public rejoicing erupted. A hundred guns were fired at New York's City Hall at dawn and again at noon. Flags flew everywhere. That night a great fireworks display was launched from the roof of City Hall, with unfortunate results. "Here in New York," George Templeton Strong wrote in his diary, "the triumphant pyrotechnics with which our city fathers celebrated this final and

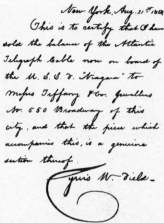

New York, Aug. 21st 1858

This is to certify that I have sold the balance of the Atlantic Telegraph Cable now on board of the U. S. S. F. "Niagara" to Messrs Tiffany & Co. Jewellers No. 550 Broadway of this city, and that the piece which accompanies this, is a genuine section thereof.

Cyrus W. Field.

A piece of the cable sold to Tiffany and Company.

Letter by Field attesting to the authenticity of the cable.

complete subjugation by man of all the powers of nature—space and time included—set City Hall on fire, burned up its cupola and half its roof, and came near destroying the County Clerk's office and unsettling the titles to half the property in the city."34*

The next day the *Niagara* arrived in New York bearing the hero of the hour, Cyrus Field, and the remaining part of the cable. Field promptly sold this to Tiffany and Company, which cut it up into short lengths, bound them in brass, and sold them as souvenirs at a handsome profit.

In all the hoopla, however, what the public wasn't told was that Queen Victoria's ninety-nine-word message had taken sixteen and

*Happily, the building—one of the masterpieces of early-nineteenth-century American architecture—was restored and it remains the seat of government for the City of New York.

a half hours to be transmitted, a rate that averaged ten minutes per word. The fact of the matter was that the cable was not working well at all. On August 11, the entire day's traffic had consisted of the following:

> Repeat, please.
> Please send slower for the present.
> How?
> How do you receive?
> Send slower.
> Please send slower.
> How do you receive?
> Please say if you can read this.
> Can you read this?
> Yes.
> How are signals?
> Do you receive?
> Please send something.
> Please send V's and B's.
> How are signals?

Still, the day after Queen Victoria's message was delivered, the first commercial message was transmitted to Europe, regarding a minor collision involving a Cunard liner: "Mr. Cunard wishes telegraph McIver Europa collision Arabia. Put into St. Johns. No lives lost."

On August 23, the first news dispatch came over the cable: "Emperor of France returned to Paris Saturday. King of Prussia too ill to visit Queen Victoria. Her majesty returns to England August 31st. Settlement of Chinese question. Chinese empire opens to trade; Christian religion allowed; foreign diplomatic agents admitted; indemnity to England and France. Gwalior insurgent army broken up. All India becoming tranquil."35 Although the news from China, the Treaty of Tientsin, was two months old, the news from Europe was that day's news. The world had shrunk by an order of magnitude.

On Wednesday, September 1, Cyrus Field received an official hero's welcome in New York City, with services at Trinity Church and a parade up Broadway to the outskirts of the city at Forty-second Street. He rode with the mayor in a carriage drawn by six white horses. The next night, he was tendered a banquet at the city's foremost hotel, the Metropolitan, at Broadway and Prince Street, featuring, in typical Victorian style, seven courses, twenty entrées, forty desserts, and endless speeches.

But that afternoon he had been told by de Sauty, who had remained in Trinity Bay, that the signals coming through the cable were so weak as to be nearly unintelligible. At the banquet itself, Field had received a message saying, "C. W. Field, New York: Please inform . . . government we are now in position to do best to forward—" The rest was blank. The cable was dead.

FOR THE NEXT FEW WEEKS, while the company did its best to conceal matters, the electricians did their best to revive it. But there was little they could do beyond adjusting batteries, trying different signal strengths, and hoping against hope.

A few feeble signals were received, but nothing more, and finally the company had no choice but to admit the situation. The public jubilation quickly turned to scorn. The *Boston Courier* claimed that the whole thing had been merely a stock swindle to allow Field to unload worthless shares on the public.*

The newspapers, which had fallen all over themselves to lionize Field and his technicians, now lampooned them. C. V. de Sauty, who had missed the brief glory days by staying at Trinity Bay, was ridiculed for his name:

> Thou operator, silent, glum,
>> Why wilt thou act so naughty?
> Do tell us what your name is—come!
>> De Santy or De Sauty?
> Don't think to humbug any more,
>> Shut up there in your shanty,
> But solve the problem once for all—
>> De Sauty or De Santy.[36]

There were even claims that the entire undertaking had been a fraud, that there was no Atlantic cable and Queen Victoria's message had actually been sent by ship.† These rumors were sufficiently widespread that Matthew Field devoted an entire chapter of his history of the Atlantic cable to demonstrating that the actual existence and brief functioning of the cable were beyond doubt.

By the end of the year, when all hope of the cable's coming back to life had long ended, many old friends and partners in the enterprise were avoiding Field. Only Peter Cooper was as staunch

---

*In fact, in the month of August 1858, Field sold exactly one share of stock in the Atlantic Telegraph Company, taking a loss of $500.

†A hundred and eleven years later, there were persistent rumors that the landing on the moon had been bogus and that the whole thing had been elaborately staged in a New York television studio.

as ever. "We will go on,"[37] he told his friend, partner, and neighbor.

Cooper may have been simply displaying the indomitable optimism that had taken him, as well as Cyrus Field, so far in life. But there was already evidence that he was correct. On August 27, the British government, realizing that the Indian Mutiny was nearly over, had used the cable to countermand orders for two regiments to board ship at Halifax, Nova Scotia, and sail for India. By carrying those new orders, the cable had saved the government between £50,000 and £60,000, more than one-seventh the total investment that now lay useless beneath the Atlantic. With such potential to save money, there could be no doubt that the desire of the British government to have a functioning Atlantic cable had been only whetted.

IX

"HERE'S THE SHIP TO LAY
YOUR CABLE, MR. FIELD"

THE LATE 1850S AND EARLY 1860S were not a good time for sub-marine telegraphy. Not only had the Atlantic cable failed after only a few weeks of fitful operation but the cable laid in the Red Sea the following year as the first segment of a line to India had likewise failed. In fact, by 1861, although 11,364 miles of submarine cables had been laid worldwide, only about 3,000 miles were still operating. The debacle of the Indian cable had cost £800,000, and the British government, anxious to have instant communica-tion with its Indian Empire since the mutiny of 1857–58, had guaranteed the interest on that sum regardless of the results. That meant that the Exchequer would be paying £35,000 a year for twenty years for nothing.

With these two spectacular failures, there were loud calls for a commission of inquiry and one was set up to hear evidence, with four members appointed by the Board of Trade and four appointed by the Atlantic Telegraph Company. They were a distinguished group, including Professor (later Sir) Charles Wheatstone, one of the earliest pioneers of telegraphy. Together with William Cooke, he had taken out a patent on an electric telegraph as early as 1837, and had helped develop numerous other electrical devices, including Wheatstone's bridge for measuring electrical resistance.* Another member was George Parker Bidder, a distinguished engineer who, however, is today remembered chiefly for his astounding ability to perform mathematical calculations in his head.†

The committee's report, somewhat longer than the Bible, drew numerous conclusions regarding what had gone wrong in submarine telegraphy up to that point. It led to significant advances both in submarine telegraphy and electrical engineering in general.

One point made was that there were as yet no agreed-upon units with which to measure such matters of fundamental importance as current and resistance. Even these terms were not yet standardized, and such vague terms as *intensity* and *tension* were used as well. This made it nearly impossible for a scientific literature on electricity to develop. Not long after the report was published, Sir Charles Bright and Latimer Clark submitted a paper to the British Association for the Advancement of Science, suggesting a systematic basis for electrical testing, employing a precise vocabulary.

In the next few years, such terms as *watt, volt, ohm,* and *ampere* came into use, each rigorously defined. The terms were based on the names of early pioneers in science and engineering, the British James Watt (1736–1819), the Italian Alessandro Volta (1745–1827), the German Georg Ohm (1787–1854), and the French André-

---

*He also invented the concertina, and helped develop the stereoscope, the immensely popular Victorian parlor entertainment. For his "sound magnifier" he coined the word *microphone.*

†At the age of ten he had been asked how many times a wheel five feet ten inches in diameter would revolve on a journey of 800,000,000 miles. In less than a minute he gave the correct answer, "724,114, 285, 704 times with twenty inches left over." This capacity never left him. Because Bidder—unlike many lightning calculators—was a highly intelligent and educated man, he was able to give valuable insights into how he performed his calculating feats, information that has been utilized by psychologists ever since.

Marie Ampère (1775–1836). It was the first instance of the scientific community honoring its own in this way.

A second point was that it was necessary to test thoroughly the cable before it was laid, not only for electrical conductivity, but also for the integrity of the insulation. In laying the cable from the Persian Gulf to India in 1863–64, Bright and Clark developed a system for subjecting the cable to hydraulic pressure equal to what it would experience when laid, in order to test it under realistic conditions.

With regard to the Atlantic cable, the conclusion was inescapable: the Atlantic Telegraph Company had been much too hasty in attempting to manufacture and lay a cable in 1857. This could be laid at the feet of Cyrus Field, whose impetuosity and drive were at once his greatest virtues and his worst defects as an entrepreneur. Dr. Edward Whitehouse, the company electrician, testified that Field wouldn't let him have the time necessary for some experiments. "Mr. Field," he told the committee, "was the most active man in the enterprise, and he had so much steam that he could not wait so long as three months. He said, 'Pooh, nonsense, why, the whole thing will be stopped; the scheme will be put back a twelve-month.'"[1]

But the main blame fell on Dr. Whitehouse himself, who had been fired even before the cable had failed completely. The directors had explained his dismissal in a letter to Thomson, referring to the recipient in the third person for some reason.

Mr. Whitehouse has been engaged some eighteen months in investigations which have cost some £12,000 to the company and now, when we have laid our cable and the whole world is looking on with impatience. . . . [W]e are, after all, only saved from being a laughing-stock because the directors are fortunate enough to have an illustrious colleague [Professor Thomson] . . . whose inventions produced in *his own* study—at small expense—and from *his own resources* are available to supersede the useless apparatus prepared at great labor and enormous cost. . . . Mr. Whitehouse has run counter to the wishes of the directors on a great many occasions—disobeyed time after time their positive instructions—thrown obstacles in the way of everyone, and acted in every way as if his own fame and self-importance were the only points of consequence to be considered.[2]

Whitehouse, in Valentia, had disconnected Thomson's exquisitely sensitive mirror galvanometer and hooked up his own patented relay device, which was supposed to write down the messages automatically. He had also insisted on using his giant induction coils, which were capable of generating thousands of volts, instead of the much more modest voltage of batteries that could be read by the mirror galvanometer. It is the best guess, and it can never be more than that, that the ultimate failure of the cable was due to the fact that Dr. Whitehouse burned it out with voltage beyond what it was capable of carrying.

The problem with his relay device was, simply enough, that it didn't work. Professor Thomson testified that "I find altogether two or three words and a few more letters that are legible, but the longest word which I find correctly given is the word 'be.'"[3]

Dr. Whitehouse defended himself furiously at the hearings, producing clouds of experimental data to justify his theories and refusing to admit that he had erred in any significant way. How well his arguments fared in the court of history may be judged by the fact that today he is utterly forgotten while William Thomson, Lord Kelvin, lies in Westminster Abbey.

When the hearings finally ended, the committee had no hesitation in its conclusions. "The failures of the existing submarine lines which we have described have been due to causes which might have been guarded against had adequate preliminary investigation been made into the question. And we are convinced that if regard be had to the principles we have enunciated in devising, manufacturing, laying and maintaining submarine cables, this class of enterprise may prove as successful as it has hitherto been disastrous."4*

Determining what went wrong, however, and getting investors to risk their money on another attempt to lay the Atlantic cable were two very different things. As the secretary of the Atlantic Telegraph Company testified at the hearings, "I myself have personally waited upon nearly every capitalist and mercantile house of standing in Glasgow and in Liverpool, and some of the Directors have gone round with me in London for the same purpose. We have no doubt induced a great many persons to subscribe, but they do so as they would to a charity, and in sums of corresponding amount."5

As THE DIRECTORS of the Atlantic Telegraph Company had been struggling to lay the cable and then to raise money for a third attempt, another product of mid-nineteenth-century technology that would turn out to be crucial to final success had been astonishing the world. It was a ship named the *Great Eastern,* and she was far more than twice as large as the *Agamemnon* and the *Niagara* combined.

There is a reason that ships, alone among inanimate things, are

---

*Committees such as this one have frequently been employed since, following other technological calamities. The *Titanic* disaster in 1912, the early crashes of the De Havilland *Comet*—the first commercial jet airliner—in the early 1950s, and the *Challenger* explosion in 1986, are three examples in the twentieth century. Such inquiries have proved a very powerful tool in determining not only what went wrong but also how to do things right in the future.

often accorded a personal pronoun in the English language. Oceangoing vessels are almost always built one at a time, often with a wholly or partially unique design, and thus each is as idiosyncratic as an individual human being. And each vessel inevitably has a unique history, a life story so to speak, just as humans do, histories that can be as filled with triumph and tragedy as our own.

Perhaps no ship in human history was so nearly unique as the *Great Eastern*. By the time she was laid down in 1857 on the Isle of Dogs, across the River Thames from Greenwich, east of London, steamships had taken over the transatlantic passenger business from the old sailing packets. But sailing ships, far less expensive to operate, still carried the world's cargo. Only in the twentieth century would the majority of the world's ocean freight move by steam.

But, unassisted, steamships could not yet reach India, the Far East, or Australia in the 1850s. The reason was fuel capacity. Coal at that time was mined on an industrial scale only in Western Europe and the United States. And no vessel was capable of carrying enough to reach halfway around the globe under steam power alone without refueling. So, for a ship to steam to Australia, coal had to be expensively carried by sailing ships, called colliers, ahead of time to such way stations as Cape Town, South Africa, and Ceylon. The greatest engineer of the nineteenth century, an Englishman named Isambard Kingdom Brunel, wanted to change that.

Brunel came from an engineering family. His father, Marc Isambard Brunel, was born in 1769 in the French province of Normandy, where his parents held a substantial tenant farm and operated a coach station. A younger son, Marc was destined for the church but rebelled against that fate and joined the French navy instead, where his antipathy to classical studies was no defect and his passion for drawing, mechanics, and mathematics a great

Isambard Kingdom Brunel

advantage. After six years he left the navy and found himself in Rouen, a royalist stronghold, as the Revolution raged in Paris. In Rouen he met Sophia Kingdom, an English girl who had come to the city to perfect her French.

As the political situation in revolutionary France deteriorated rapidly, Brunel decided to flee to America and spent six years in New York, becoming both an American citizen and chief engineer of the port. In 1798 a chance encounter at dinner with another French émigré changed his life.* He learned that ships' blocks for the Royal Navy were made under an exclusive contract by a firm called Messrs Taylor of Southampton. Ships' blocks are the

---

*There were about 25,000 French émigrés in the United States in the 1790s, including Talleyrand, who returned to France to become foreign minister under the Directory and Napoleon.

wooden (or, now, often metal) casings that hold one or more pulleys. Used in the running rigging to allow the sailors, using the pulleys' mechanical advantage, to trim the immensely heavy yards, these blocks are vital to the working of a sailing ship.

A seventy-four-gun ship of the line of that time needed no fewer than 1,400 of them. With the Royal Navy maintaining a fleet of hundreds of ships on active duty in the war against France, it purchased more than 100,000 a year. The cost of ships' blocks was no small portion of the navy's budget.

Brunel saw an opportunity and quickly devised greatly improved machinery for manufacturing ships' blocks. He sailed to England, where he sought out and married Sophia Kingdom and presented his ideas to the admiralty. When finally constructed — no small task, as it pushed the limits of mechanical engineering — his machinery was a great success, able to manufacture blocks with only one-tenth of the personnel previously required, and thus effectively at one-tenth the cost. The Royal Navy paid Brunel £17,000, a modest fortune by the standards of the day, for his work. This work so impressed Czar Alexander I of Russia, who visited the shipyard, that Alexander presented Brunel with a large ruby set with diamonds and gave him a standing invitation to come to Russia. Before his machines were even up and running, his wife presented him with his only son, Isambard Kingdom Brunel, born in 1806.

Marc Brunel soon became one of Britain's most noted engineers, and his son showed every sign of following in his father's footsteps. He demonstrated a marked talent for drawing at four and had mastered plane geometry by the time he was six. While still a boy Isambard demonstrated the greatest gift an engineer can possess, an intuitive sense of what is structurally sound. When a new building was under construction across the street from his school, Isambard predicted that it would collapse, and it soon did.

His father saw to it that he received a first-class education in engineering, both in Britain and in France, after the Napoleonic Wars ended, where he attended the Lycée Henri-Quatre, then famous for its mathematics instruction. At the age of seventeen he went to work in his father's office. By the time he was only twenty-four, his reputation was such that he was elected a fellow of the Royal Society.

Marc Brunel's greatest project was the construction of a pedestrian tunnel under the Thames, begun in 1825, a project on which his son was soon named resident engineer. There had been a previous, disastrous attempt to tunnel under the Thames in the early decades of the nineteenth century. But there was no successful precedent for a large underwater tunnel, other than the extension of some of the Cornish tin mines out under the seabed.* To accomplish the task, the elder Brunel designed his most famous invention, the tunneling shield, patented in 1818. It has been the basic technology of deep tunnel construction ever since. The idea had come to Brunel when he noticed the armored heads of teredo worms, a marine pest that plagued wooden ships by boring into their hulls in great numbers.

The tunnel, located east of Tower Bridge (and now long part of the London subway system), would take eighteen years to complete, with numerous disasters, both financial and structural, in the course of construction. Indeed, the younger Brunel was very nearly killed when one wall of the tunnel was breached under pressure from the water outside. The tunnel rapidly filled with water and

---

*The need to find an efficient means to pump out the water that inevitably leaked into these mines prompted Thomas Newcomen to devise the first practical steam engine in 1712. James Watt's radical improvement of Newcomen's engine, patented in 1769, then sparked the Industrial Revolution.

Brunel was trapped by a fallen timber. By a great effort he freed himself and then had the good fortune to catch a wave that shot him up the shaft of the visitors' stairway, allowing Brunel to escape the fate of six other workers in the tunnel, who drowned.

Like his father, Isambard Brunel was the master of all forms of engineering, from docks, to suspension bridges, to water towers, to prefabricated field hospitals, to tunnels, to railways, to ship design. His was the last generation of engineers who could spread themselves so wide, and Brunel proved himself a genius in each of these engineering fields. In 1910, the magazine *The Engineer* said of him, simply, that "in all that constitutes an engineer in the highest, fullest and best sense, Brunel had no contemporary, no predecessor. If he has no successor, let it be remembered that . . . the conditions which call such men into being no longer have any existence."[6]

His Great Western Railway, which ran originally from London's Paddington Station (which he designed, of course) to Bristol, is considered his masterpiece. He designed almost every aspect of it. The Box Tunnel, in Gloucestershire, the longest in the world at the time at 1⅞ miles, was considered a wonder and is a perfect example of the extraordinarily close attention Brunel paid to every aspect of his work. Taking advantage of its fortuitous geographic orientation, he deliberately sited the tunnel precisely so that the rising sun shone along its entire length on April 9 every year. April 9 was Brunel's birthday.

Such touches also reveal Brunel's great weakness: a near total inability to delegate, which led him to drive himself relentlessly to the point of exhaustion. During a crisis in the construction of the Thames Tunnel, he worked for ninety-six hours straight. In the midst of construction of the Great Western Railway, Brunel wrote that "it is harder work than I like. I am rarely much under twenty hours a day at it."[7]

The Great Western Railway led Brunel into ship design. The port of Bristol had been losing business to Liverpool, which was much closer to the industrial centers of Birmingham and Manchester, which exported to the world. Local merchants hoped to use the railroad, with its quick access to London, to capture the transatlantic passenger business. They intended, in effect, to extend the Great Western Railway clear across the Atlantic by means of steamships, a revolutionary concept in the 1830s.

Although the *Savannah* had first crossed the Atlantic in 1819, her engine was only auxiliary, designed for use in calm weather and adverse winds. Many felt that an all-steam crossing of the Atlantic from England to New York was inherently impossible. The science writer and mathematician Dionysius Lardner thought that people "might as well talk of making a voyage from New York or Liverpool to the moon."[8]

Brunel knew better, and in 1838 designed a 1,340-ton ship called the *Great Western*, the first purpose-built transatlantic steamer and by far the largest steamer yet constructed. Built of wood but with many iron elements, the *Great Western* was highly innovative, as might be expected of Brunel. On her maiden voyage she smashed the transatlantic speed record, crossing in only fifteen days and five hours. From then on, transatlantic passenger ships carried sails only for use in the event (a not infrequent one, to be sure) that the engines broke down.

Like many engineers at the time, notably the Swedish John Ericsson, who designed the USS *Monitor* in 1861 and thereby revolutionized naval warfare, Brunel was soon deeply interested in screw propulsion. The screw had several advantages over paddle wheels. Being safely tucked below the waterline, it was much less subject to damage from the vagaries of wind, wave, or enemy action. And because a screw is cast as a solid piece of bronze —

still the metal of choice—it was much less fragile than the large wooden paddle wheels. At the time, however, a screw, needing a much higher rpm than paddle wheels to produce the same forward motion, put great demands on the design of the engines that drove it.

In 1843 Brunel's second ship, the *Great Britain*, was launched by Prince Albert. Iron hulled and screw driven, it was in many ways the first modern passenger ship and the first such ship to cross the Atlantic. She also clearly showed the possibilities for size that iron made possible. With a length of 322 feet, a beam of 50 feet 6 inches, and a displacement of 3,500 tons, she was not only larger than any steamship built before, she was larger than any purely wooden ship could be.*

With iron construction, there was little physical limit to the size of a ship. And a decade after Brunel conceived of the *Great Britain*, great size was suddenly more fashionable than ever. Characteristically, he seized the moment and created the most extraordinary ship the world has ever known.

IN 1851, THE GREAT EXHIBITION held in the Crystal Palace—the first world's fair, organized by Prince Albert—was opened in Hyde Park in London. It electrified the imagination of the world with its display of the fruits of the Industrial Revolution. The

---

*The *Great Britain* ended her days when she was severely damaged in a gale off Cape Horn in 1886. Condemned as unseaworthy, she was used as a store ship until 1933 in the Falkland Islands. Afterward, she was beached and allowed to rot until 1970, when she was brought back to Bristol, the city of her birth, and restored. She may be seen, and visited, today in the dock where she was built.

devices and manufactures from many countries brought 6.2 million visitors, who paid a shilling and up to see them.†

But the Crystal Palace itself was the biggest draw. Designed by Sir Joseph Paxton, a garden designer who had built a very large greenhouse for the Duke of Devonshire, it was constructed of cast- and wrought-iron structural members and 294,000 panes of glass. At 1,550 feet long and 450 feet wide, it was no less than three times the length of St. Paul's Cathedral and yet was constructed in only nine months, thanks to its prefabricated parts. Also thanks to these parts, this enormous building, after the Great Exhibition closed, was easily dismantled and re-erected in Sydenham, in South London, where it was the center of an amusement park until it was destroyed by fire in 1936.

The Crystal Palace was as influential as it was innovative. The possibilities of both scale and scope opened up by the Industrial Revolution were manifest, and thinking big became very much in fashion in the 1850s.

In this quintessentially Victorian atmosphere of bigger and better, Brunel, who had headed the section on civil engineering of the Great Exhibition, wrestled with the problem of steaming to the Orient. Deeply experienced in the design of unprecedentedly large vessels, he was well aware that while the energy required to drive a ship through the water increased, roughly, as the square of her dimensions, her carrying capacity increased as the cube. Therefore, if a ship were large enough, she could carry coal sufficient to steam anywhere. Brunel's genius lay in having the courage to design a ship fully five times the size of any ever built before. "Nothing is proposed," wrote Brunel, trying to attract investors,

---

†The proceeds were used to purchase land in Kensington, where today the Victoria and Albert, Science, and Natural History Museums are located.

"but to build a vessel of the size required to carry her own coals on the voyage" to Trincomalee, Ceylon, and return, a distance of 22,000 miles.9*

Brunel argued that this would be a paying proposition, because the ship would capture much of the trade between Britain and Britain's Eastern empire. Indeed, he calculated that his great ship would earn 40 percent on her investment per year. Brunel, a persuasive man as well as a great engineer, induced a group of investors to put up the then enormous sum of £600,000 to build her.

Just how colossal she was still staggers the imagination nearly 150 years after her launching. At 693 feet long with a beam of 120 feet (including the paddle wheels), she would not have been able to utilize the Panama Canal, finished more than half a century later. At 22,500 tons displacement, she held the size record for nearly half a century, until long after she had been broken up. Not until the ill-fated *Lusitania* slid down the ways in 1907 did a larger vessel than the *Great Eastern* float upon water. Throughout her life, no matter what harbor she was in, from Baltimore to Bombay, the *Great Eastern* dwarfed all other vessels, looking a bit like a rowboat in a pond full of ducks.

---

*There was to be a curious echo of the reasoning behind the design of the *Great Eastern* a hundred years later. During World War II, the U.S. Army Air Corps wanted the capacity to transport troops in large numbers overseas by air, protecting them from submarine attack. It commissioned Hughes Aircraft Company and Henry Kaiser to design a craft capable of carrying large numbers of troops. Kaiser soon dropped out of the project, but the obsessive Howard Hughes continued even after the contract had been canceled with the end of the war. The result was a seaplane built of wood (whence its nickname, the *Spruce Goose*) that was, for decades, the largest airplane ever constructed. It still holds the record for number of piston engines (eight) and wingspan. At nearly 320 feet, the wingspan of the *Spruce Goose* exceeds that of a Boeing 747 by more than one hundred feet. The plane flew only once, for about 1,000 yards. Today it is on exhibit in McMinnville, Oregon.

The Great Eastern (second from the bottom) compared in approximate scale to (from top to bottom) an average 1800s sailing ship, the Great Western, *the* Great Britain, *and a typical ocean liner built nearly a hundred years after the* Great Eastern.

Almost every aspect of her was an innovation or a record. She was the first ship built without ribs, deriving her structural strength from a double iron hull. She featured the first watertight compartments, the first powered steering gear. So great was the demand she created for wrought-iron plates — she needed 30,000 of them, each averaging a third of a ton in weight — that it significantly drove up the cost of iron on the world market.

Although the River Thames is about a thousand feet wide at Greenwich, the *Great Eastern* had to be built parallel to the river and launched sideways—the first major ship so launched. Had she been launched stern first, her nearly seven-hundred-foot length would have made it impossible to check her way in such a narrow space in time to avoid the opposite shore.

The engines were, by far, the largest and most powerful ever built to that date. And although Brunel had been the first to abandon the fragile, cumbersome, and dangerous paddle wheels in favor of screw propulsion in the *Great Britain*, he employed both paddle wheels and a single screw in the *Great Eastern*. The reason is that it was not then possible to apply sufficient power to a single engine shaft to propel so large a vessel at a good speed. And at that time engines were still too bulky to allow two of them to be placed parallel to drive twin screws.

As a result there were two completely separate power plants. The paddle wheels, which measured fifty-eight feet in diameter, were the largest ever built for a ship. They were designed to turn at an astonishing 17 rpm, powered by engines that produced up to 3,800 horsepower. The screw, which measured twenty-four feet in diameter, and resembled, one contemporary said, "the bone blades of some pre-Adamite animal,"[10] was powered by engines producing 6,200 horsepower.* The two engines together, it was said, could have powered all

---

*The blades on the *Great Eastern*'s screw were nearly flat, like the blades on a ceiling fan. Later, much more sophisticated design, involving very complex mathematics, would allow ships to have smaller screws while producing the same forward motion per revolution. This saved both considerable weight and considerable expense, for bronze is very expensive. Not until the advent of nuclear submarines—whose massive, slow-turning, multibladed screws allow extremely quiet operations and great propulsive force at the same time—would a vessel have a larger screw than the *Great Eastern*.

the cotton mills in Manchester. While that might have been hyperbole, there is no doubt that the two systems could move the *Great Eastern* at speeds approaching twelve knots, a very good speed in the 1850s.

In addition to her steam engines, the *Great Eastern* also sported no fewer than six masts (five of which served as funnels for auxiliary engines, yet another innovation). They could carry up to an acre and a third of canvas. With her five funnels (no other steamship has ever had more than four), six masts, and two paddle wheels turning once every 3.5 seconds at full speed, the *Great Eastern* was a magnificent sight as she steamed. Although she appears strange to our eyes because she lacked the superstructure that soon became standard on passenger ships, her handsome lines have long been admired by naval architects.

And, like the Atlantic cable and the Crystal Palace that in part inspired her, she fascinated the people of her day. Composers wrote popular songs in her honor. She was called "the floating city" and the "crystal palace of the sea." Currier and Ives pub-

A Currier and Ives lithograph of the Great Eastern.

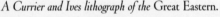

lished no fewer than five lithographs of her before she ever put to sea. Longfellow wrote a poem about her.

At her scheduled launching on November 3, 1857, thousands of spectators appeared onshore and in the river to watch the largest movable object yet constructed move for the first time. Their conversation made it nearly impossible for Brunel to communicate with those controlling the hydraulic rams (to push her) and windlasses (to check her) onshore and the steam tugs to pull her into the river.

The launch was a disaster. The ship, groaning like some vast awakening beast, began to slide down the ways, but when the windlass crews tried to check her speed, the ship shuddered to a stop and would not move again. Slowly, over the next two months, Brunel was able to get her halfway down the ways. On January 5, having assembled no fewer than twenty-one hydraulic rams, he tried again to launch her. The ship moved three inches before the rams began to burst with the effort. Finally, on January 31, 1858, with the help of a spring tide, the *Great Eastern* finally floated in the filthy water of the Thames and was towed across to the far shore to be completed. The launch from where she was built to where she floated, it is estimated, had cost £120,000, about £1,000 a foot, at a time when £1,000 was an upper-middle-class annual income.

The ship, thanks partly to her launch, had cost twice as much to build as originally estimated and drove the shipyard owner into bankruptcy. The corporation formed to finance the ship was also broke and the *Great Eastern* was sold to a new syndicate for £165,000, less than a third of her projected construction costs, before she ever sailed.

Hard luck as well as failure would stalk her throughout her life. On her shakedown cruise she suffered an explosion that demolished the grand saloon and the forward smokestack and

The Great Eastern *being launched broadside.*

killed five members of the crew. Rounding the tip of Long Island on her way to Flushing Bay, she ripped an eighty-three by nine-foot gash in her hull on a previously uncharted rock—still known as Great Eastern Rock. The accident would have sunk any other ship in the world in a few minutes, but the *Great Eastern's* inner hull held and she was able to continue on to Flushing Bay, although listing heavily to starboard.

No dry dock in the world could accommodate her, and because of her flat bottom, she could not be careened onshore to reach her bottom. A whole new technology, developed by the New York engineer Edward S. Renwick, had to be employed so that she could be repaired while afloat.*

Perhaps the hardest luck to befall the *Great Eastern* was the

---

*His brother James, an architect, designed St. Patrick's Cathedral in New York and the Smithsonian Institution in Washington, D.C.

John Scott Russell, Lord Derby, and Isambard Kingdom Brunel

fact that she killed her creator. Brunel worked himself beyond endurance as he wrestled with the endless complexities of building so revolutionary a ship. Only fifty-one when she was finally launched, his friends noticed how he had aged over the course of the project. His once thick brown hair had mostly fallen out. Always notably short (he was often called "the little giant"), he now walked with a stoop.

On the day before his great ship was to begin her sea trials, Brunel climbed painfully aboard to have a final look at her and intended to sail with her. Standing before the mainmast, he was photographed with Scott Russell, in whose shipyard she had been built, and Lord Derby, who had been prime minister until a few months earlier. Hours later, Brunel collapsed with a stroke. Although his mind was clear, he was paralyzed. A week later, shortly after hearing of the explosion in her forward funnel, he died.

But while the *Great Eastern* was still on the ways and the hopes and dreams of her owners and builder had not yet been dashed upon the rocks of economic reality, Brunel had his last great engineering insight.

Cyrus Field had met Brunel by chance on a railroad returning from Valentia to London. Brunel instantly saw the synergistic possibilities in their two dreams and invited Field to visit the yard at the Isle of Dogs to see the great ship under construction. As Brunel showed Field around, the engineer told him, "Here is the ship to lay your cable, Mr. Field."[11]

Indeed she was. It is by no means the least of her ironies that the one thing the *Great Eastern* turned out to be good for—indeed perfect for—was perhaps the only use her creator had not foreseen from the beginning.

X

A NEW CABLE, A NEW ATTEMPT

BY THE TIME THE COMMITTEE had released its report, the American Civil War was raging across half a continent, greatly complicating any attempt to lay the cable. The Civil War was by far the greatest war fought in the Western world in the nineteenth century. Its size was unprecedented, with a front ranging from Virginia to Texas and with armies numbering in the millions. So were its costs, both in money (the American national debt rose from $65 million to $2.8 billion) and in human lives (nearly 3 percent of the male population died in uniform over the four years it lasted).

But wars are always costly in both treasure and lives. What made the American Civil War so unprecedented was that it was the first great war of the industrial era. This allowed not only greatly increased production of the matériel of war but a revolution in command and control as well. Railroads and steamboats made possible the rapid movement of large numbers of troops, and the telegraph enabled the entire war to be directed from Washington and Richmond in real time. When President Lincoln, on April 15, 1861, called for 75,000 volunteers after Confederate forces attacked Fort Sumter in Charleston harbor, the call reached the most distant parts of the Union states almost immediately. The extraordinary reaction—an outpouring of men, and offers of help from bankers, manufacturers, and wholesalers—powerfully affected public opinion in the North, a public opinion that reinforced itself as it was communicated through the telegraph system.

The war also greatly complicated relations with the United Kingdom. Heavily dependent on imports of Southern cotton to keep its vast textile industry running, Britain was sympathetic to the Confederacy. It quickly declared neutrality and granted the South the status of a belligerent power. It also turned a blind eye to the construction of Confederate commerce raiders, most notably the CSS *Alabama*, in British yards.*

The most serious dispute between Britain and the United States, the so-called Trent affair, led to the brink of war. Captain Charles Wilkes, commanding the USS *San Jacinto*, had stopped the British merchant ship *Trent* in international waters and removed two Confederate agents. When the news of this flagrant violation of international law reached Britain, the nation exploded in anger and the government ordered troops to Canada and naval preparations. Lincoln, quickly deciding on "one war at a time,"[1] ordered the release of the agents and disavowed Wilkes's action.†

The seizure of the Confederate agents occurred on November

---

*This would lead to the *Alabama* arbitration in 1871 to settle the dispute between Britain and the United States over claims for compensation resulting from the actions of the CSS *Alabama*. The arbitration had an enormous impact on the development of international law and become a powerful precedent for settling international disputes peacefully.

†In the *Trent* affair, Prince Albert performed his last great service for the country that had never learned to appreciate or even like him. He toned down the ultimatum written by Earl Russell, the foreign minister, and provided an out for the American government by including the suggestion that Wilkes had been acting without orders, which, in fact, he had been. Already mortally ill, Albert died two weeks later.

8, but news of it did not reach England until November 27. Prince Albert's memo was drafted on December 1, and the American government released the agents shortly after Christmas. Everyone was aware that if the Atlantic cable had been in place, the entire crisis would have played out very differently, although no one was sure how. The young Henry Adams, serving as an unpaid secretary to his father, Charles Francis Adams, U.S. minister to Great Britain, thought that the cable would have made war certain, because there would not have been time for tempers to cool. *The Times* of London, however, wrote that "we nearly went to war with America because we had not a telegraph across the Atlantic."[2]

Cyrus Field moved at once to exploit the situation. He wrote to George Saward, secretary of the Atlantic Telegraph Company, that "now is the time . . . to act with energy and decision, and get whatever guarantee necessary from the English government to raise the capital to manufacture and lay down without unnecessary delay between Newfoundland and Ireland a good cable."[3]

He wrote to William Seward, Lincoln's secretary of state, "The importance of the early completion of the Atlantic telegraph can hardly be estimated. What would have been its value to the English and United States governments if it had been in operation on the 30th of November . . . ? A few short messages between the two governments and all would have been satisfactorily explained. I have no doubt that the English government has expended more money during the last thirty days in preparation for war with this country than the whole cost of manufacturing and laying a good cable between Newfoundland and Ireland."[4]

Seward met with Field and wrote Charles Francis Adams in London, instructing him to seek action from the British government on the cable. Field sailed again for England and met with Earl Russell and Lord Palmerston, the prime minister. Palmerston asked numerous questions but made no specific promises beyond agreeing to increase the guarantee minimum government cable usage to £20,000 a year, provided, of course, that the cable worked.

Field then embarked on a public relations campaign in hopes of reviving interest in the cable, including holding a "telegraphic soiree" at the house of Samuel Gurney, one of the directors of the Atlantic Telegraph, at Prince's Gate, near Hyde Park. The purpose was "to spread information on the subject among all classes, but chiefly among educated and intelligent persons."[5] Wires were brought in connecting the house to every telegraph system in Europe, and the guests viewed various types of telegraph equipment and cables and were treated to demonstrations. The earl of Shaftesbury, the great social reformer, sent a message to St. Petersburg inquiring about the health of the czar, and in four minutes back came the news that the czar was well.

Back in the United States after yet another ten-day voyage, Field traveled widely trying to interest investors. But raising money was a very slow process after the great failure of 1858. Henry Field remembered his brother's description of a meeting he had with "the solid men of Boston." They "listened with an attention that was most flattering to the pride of the speaker, addressing such an assemblage in the capital of his native state. There was no mistaking the interest they felt in the subject. They went still farther, they passed a series of resolutions, in which they applauded the projected telegraph across the ocean as one of the grandest enterprises ever undertaken by man, which they proudly commended to the confidence and support of the American public, after which they went home, feeling that they had done the generous thing in bestowing upon it such a mark of their approbation. *But not a man subscribed a dollar!*"[6]

"Other cities," Henry Field wrote, "were equally gracious, equally complimentary, but equally prudent."7 Only in New York, where Cyrus Field was known personally by most of the men of means and very well liked, did he find it possible to raise money for a new attempt. Altogether, Field raised about $330,000 in New York, but "even those who subscribed," his brother wrote, "did so more from sympathy and admiration of his indomitable spirit than from confidence in the success of the enterprise."8

Field's own finances were now precarious. In 1860 he had suffered a disastrous fire at his warehouse. His losses amounted to $120,000, only two-thirds covered by insurance. By the end of that gloomy year, as the Union unraveled and the American economy remained mired in depression, he figured that he had lost a total of $130,000. Once more he asked his creditors to accept a settlement, paying them twenty-five cents on the dollar until better times arrived. He mortgaged nearly everything he owned, including his furniture and his pew at the Madison Avenue Presbyterian Church.

At least that year saw the consolidation of his interests in domestic telegraph lines with those of Samuel Morse. In the very early days of telegraphy, there were hundreds of separate companies, whose lines did not interconnect. Consolidation was inevitable and Peter Cooper, together with Field and others involved in the cable, had formed the American Telegraph Company to further this. Morse had at first greatly resented it and there was a breach in the friendship between Morse and Field. But in 1860, Morse agreed to align his interests with the American Telegraph Company, which became dominant in the eastern United States.

That year also, Field brokered a deal with the Associated Press, one of the American Telegraph Company's most important customers, to meet ships off Cape Race, Newfoundland, and use the facilities of the New York, Newfoundland, and London Telegraph Company to get the news to New York two days sooner than the ships could carry it there themselves. It was the original conception of Frederick Gisborne in action.

The Civil War ended the depression of the late 1850s, and Field was able to pay off his creditors and soon decided to liquidate the paper firm so that he could devote himself full time to the cable and his telegraph interests. All his eggs were now in the basket of the Atlantic cable.

He did better in England, where the need for an Atlantic cable was more obvious and where the technology was more familiar. The report of the Committee of Inquiry, which was published in July 1863, made it clear that a cable was technologically well within reach. And the American Civil War had demonstrated that Britain needed fast communications with the United States and its American possessions, especially Canada. The directors of the Atlantic Telegraph Company by this point had raised £220,000. However, even with what Field had managed to raise in the United States, he still had less than half the sum needed. Matters were at a standstill.

But, as Matthew Field wrote, "It is a long night which has no morning."9 In January 1864, Cyrus Field returned to England once more, his thirty-first trip across the Atlantic, and there he met Thomas Brassey.

Brassey, born in 1805, had begun life apprenticed to a surveyor. As he reached manhood, the railroad was becoming a practical technology and Brassey became a railroad contractor, building a portion of the Grand Junction Railway in 1835. He proved himself a master of dealing with enterprises of once unimaginable size. The London and Southampton Railroad involved capital of no less than four million pounds and a workforce of three thousand.

He was soon operating on a global scale and at one time was busy with projects in Europe, India, Australia, and South America, employing nearly seventy-five thousand men. By the 1860s he was immensely rich.

Armed with an introduction, Field called on Brassey and asked him to invest. Brassey was not an easy sell. "In attempting to enlist him in our cause," Field remembered, "he put me through such a cross-examination as I had never before experienced. I thought I was in the witness-box. He inquired of me the practicability of the scheme — what it would pay, and everything connected with it; but before I left him, I had the pleasure of hearing him say that it was a great national enterprise that ought to be carried out, and, he added, I [Brassey] will be one of ten to find the money required for it. From that day to this he has never hesitated about it, and when I mention his name, you will know him as a man whose word is as good as his bond, and as for his bond, there is no better in England."[10]

Brassey's wealth, prestige, and reputation allowed him to do what Peter Cooper had done ten years earlier when the enterprise had been in its infancy: give Cyrus Field the credibility he needed to raise money from others. "I met with a rich friend from Manchester," Field reported, "Mr. [later Sir] John Pender, and I asked him if he would second Mr. Brassey, and walked with him from 28 Pall Mall to the House of Commons, of which he was a member. Before we reached the House, he expressed his willingness to do so to an equal amount."[11]

In April 1864, the Gutta Percha Company, which had man-ufactured the core of the first cable, merged with Glass, Elliott & Company, which had provided the sheathing of that cable. The new firm, the Telegraph Construction and Maintenance Com-pany, would dominate the field of submarine telegraph manufac-ture for the next century, turning out about 90 percent of all the cable laid in that time. The new company quickly arranged to supply the rest of the required capital, £315,000, and agreed to take shares in the Atlantic Telegraph Company, once the cable was working, as their profit.

One of the directors of the Telegraph Construction and Maintenance Company was Daniel Gooch, chairman of the board of the Great Western Railway.* Gooch formed a syndicate that was the third owner of the *Great Eastern*, purchasing the vessel for a mere £25,000. Gooch, like Brunel, thought that the Atlantic cable and the *Great Eastern* "seemed to have been made for one another."[12] Gooch offered to provide the ship, which was conspicuously idle in any event, for nothing, provided that, upon success, the syndicate receive £50,000 in Atlantic Telegraph stock. Field was delighted to accept, saying, "In all my business experience I have never known an offer more honorable."[13]

AFTER SIX LONG YEARS, since the failure of 1858, the financing was now once more in place. All that remained was to design and build 2,600 miles of new cable and have cable tanks built into the *Great Eastern*'s innards.

Much had been learned since the first cable had been designed in such haste in 1857. With Dr. Whitehouse no longer involved, William Thomson and Charles Bright were the dominant figures in its design. The core of the new cable once again consisted of seven strands of copper wire twisted together, but it was almost

---

*In 1842, at the age of twenty-one, when Gooch was locomotive superintendent of the railroad, he and Brunel had been at the controls when Queen Victoria rode on a train for the first time, traveling from Windsor to London and putting the royal stamp of approval on the revolutionary new form of transportation.

three times as heavy, weighing 300 pounds to the mile, instead of 107. And the copper was to be as pure as possible to ensure good conductivity. Further, the core was subjected to hydraulic tests to make certain that it would work in conditions to be found at the bottom of the Atlantic.*

Surrounding the core were to be four, not three, layers of gutta-percha, and this time the gutta-percha would be supplemented with a new substance, called Chatterton's compound, after its inventor, John Chatterton, an employee of the Telegraph Construction and Maintenance Company. This was applied over the wires and between the layers of gutta-percha. The insulation of the new cable weighed 400 pounds to the mile, where the old cable's insulation had weighed only 261 pounds.

The armor for the cable was also increased. The technology of wire manufacture had improved in the years since the first cable had been manufactured. In 1858 Sir Henry Bessemer had introduced his process for making steel, greatly reducing the price of what had previously been a luxury metal whose use was restricted to such high-value items as sword blades and razors. Now a form of steel called charcoal iron was cheap enough to be used in the Atlantic cable. But first, the core was wrapped in hemp impregnated with pitch and each strand of the armor was also wrapped in pitch-soaked hemp.

The hemp served three purposes. It protected the iron wire from contact with seawater, it increased the flexibility of the cable, and it added bulk without adding much weight. The old cable had weighed a total of 2,000 pounds per mile; the new cable weighed almost twice as much, 3,575 pounds per mile. But the

---

*The water in which it was tested was at seventy-five degrees. The temperature of the much colder water of the deep Atlantic actually improved the insulating properties of the gutta-percha.

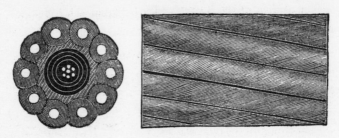

Old 1858 cable (top) and the new 1865 cable (bottom).

new cable was so much larger in circumference—1.1 inches instead of ⅜ of an inch—that it weighed much less in water than the old cable had. This, in turn, meant that it would descend into the sea at a less acute angle and be much less likely to break in consequence. Also, its breaking point was very much higher. It was estimated that ten miles of the new cable could hang vertically in water before it would snap from its own weight. As the route between Ireland and Newfoundland was no more than two and a half miles deep at any point, that would never come about.

WITH THE MONEY IN HAND and the cable design decided, Field returned once again to the United States, where the directors of the New York, Newfoundland, and London Telegraph Company hailed him as a hero, noting that "to him is due the credit, and to him this company and the world will be indebted."[14]

One matter still had to be settled on the North American side of the Atlantic: where to bring the cable ashore. Bull's Arm, where the first cable had been landed, was too far up Trinity Bay for the *Great Eastern* — which drew thirty-four feet, greater even than the largest of World War I battleships built half a century later — to reach.

Field traveled to Newfoundland once again and surveyed Trinity Bay with Captain Orlebar of the British survey ship *Margaretta Stevenson*, searching for a new spot. Field would often go ashore to reconnoiter and to gather biological specimens such as seashells and dead birds. These he packed and sent to Professor Louis Agassiz, the founder of the Harvard Museum of Comparative Zoology and the greatest American naturalist of his day. Field and Orlebar finally settled on a place on the eastern side of Trinity Bay to bring the cable ashore. No doubt it was selected for its geographic qualities of proximity to deep water and gently shoaling beach. But it is hard not to believe that Field was also attracted by its name: Heart's Content.

Field, ever restless, spent the rest of the summer traveling around the northeast, reorganizing the telegraph lines that connected with those of the New York, Newfoundland, and London Telegraph Company. He wrote George Saward in London that he had traveled by nearly every conveyance imaginable: "By railway, over 3280 miles. By steamers, over 2400. By open wagon, over 500 miles. By stage-coach, over 150 miles. By fishing-boats about 100 miles."[15]

Back in New York he held a series of dinner parties, engaged in what today would be called networking. At one party in early October, among his guests were General John A. Dix, who commanded the Department of the East, which included New York and New England, and Lord Lyons, the British minister to the United States. During dinner, General Dix was brought a message that Confederate troops, disguised as civilians, had entered Vermont from Canada, changed into uniform, and seized the town of St. Albans, fifteen miles south of the border. There they had killed one townsman, robbed the local banks of about $200,000, and set several buildings on fire.

Dix was a no-nonsense general. As secretary of the treasury in the waning days of the Buchanan administration, Dix had sent a famous telegram to a treasury agent in New Orleans after the captain of a revenue cutter had refused to turn over command of his vessel. The telegram ended, "If anyone attempts to haul down the American flag, shoot him on the spot."[16] Dix left the room and ordered federal troops stationed in Burlington, Vermont, to go after the rebels, even into Canada if necessary, "and destroy them."[17]

Dix took Field aside and told him what had happened and what he had ordered. Field was appalled. He knew that American troops entering Canada, even in pursuit of rebel soldiers, would make the *Trent* affair that had brought Britain and the United States to the brink of war three years earlier pale by comparison.

Britain was then at the height of its relative power in the world and the British public would never stand for such an insult to British sovereignty. Field immediately asked General Dix and Lord Lyons into another room to discuss the matter, and Dix soon countermanded his authorization to enter Canadian territory.

Cyrus Field was genuinely in favor of close relations between his native country and one he had grown to love. And, of course, he wanted no further delays in laying the Atlantic cable, which a war between Britain and the United States would have made

Coiling the cable in one of the tanks on board the Great Eastern.

impossible. Thanks to his quick action, what might have turned into a dark day in the history of the nineteenth century became merely a footnote, the northernmost military action of the American Civil War.

Relations between Britain and the United States were not good at this point in any event. The United States Navy refused to lend warships to assist in what would be the fourth attempt to lay the cable. It noted that the "twenty-four-hour rule," which forbade warships of belligerent powers from staying more than twenty-four hours in a neutral port, was in effect, making it impossible for American warships to assist on the west coast of Ireland.

The following March (1865), Field returned once more to England and then proceeded to Egypt. There he represented the New York Chamber of Commerce at a conference of chambers of commerce on the uses of the Suez Canal, which was nearly completed, and would, like the Atlantic cable, shrink the world enormously.

Back in England, the *Great Eastern* was making ready. With too deep a draft to reach the cable factory at Morden Wharf, she was docked at Sheerness, twenty miles downriver, and the cable was ferried to her as sections were completed. When Field returned to England on May 1, the task was nearing completion and the last mile of cable was manufactured on May 30, 1865. When it had been stowed on the *Great Eastern*, the Prince of Wales, later King Edward VII, came aboard to inspect the ship. Field invited him to send a message through the 2,700 miles of cable. "I wish success to the Atlantic cable," the prince asked to be signaled, and it was received at the other end of the cable in less than a minute.

Including the crew, there were to be five hundred people on board the *Great Eastern* for the cable laying. C. V. de Sauty was the chief electrician, assisted by William Thomson and Cromwell Varley. The chief engineer was Samuel Canning, who, long before, had been the engineer in charge of laying the cable across the Cabot Strait. The only American on board was Cyrus Field. Indeed, except for Field, the fourth attempt to lay the cable was entirely British. Only British ships were involved. Ninety percent of the capital was British, as was nearly all the technical expertise. The cable itself had been designed and manufactured in Britain.

And yet, when *The London Illustrated News* had accused Field of taking too much credit, Cromwell Varley, assistant electrician on the project, had leaped to his defense, writing, "On the contrary he seems to have given a fair share, indeed I may say the lion's share of the credit to his English associates. . . . As an Englishman I am especially grieved that my countrymen appear to give less credit to Mr. Field than is undoubtedly his due."[18]

Although many newspapers, both British and American, had applied to send correspondents along on the expedition, only one, William Howard Russell, of *The Times* of London, was granted permission to come. By this time Russell was as famous as any newspaperman in the world. With his dispatches from the field during the Crimean War, Russell had virtually invented the job of war correspondent.* Bringing the unvarnished reality of war to the breakfast tables of Britain in a way that official communiqués never could (or would), these dispatches not only helped bring down the government of Lord Aberdeen in 1855 but led to thoroughgoing reform of the British army. Russell was thus one of the most important figures in making the new mass media a major force in politics. After the Crimea, he reported from the Indian Mutiny and the American Civil War (where his early, brilliantly evocative, and highly influential dispatches earned him the nickname "Bull-Run Russell"). In later years he was one of the first journalists to be knighted.

Besides the crew and the technical staff, the *Great Eastern* carried live animals to provide food. Refrigeration was then in its infancy and the first refrigerated ship would not be built until 1869. William Howard Russell thought that "such a freight had not been seen since Noah's Ark was stranded on Mount Ararat."[19] Cyrus Field, who loved to make lists, noted that the *Great Eastern* was carrying 10 bullocks, 1 milk cow, 114 sheep, 20 pigs, 29 geese, 14 turkeys, and 500 other fowl all housed on the upper deck in an impromptu barnyard. Their assorted sounds added a strangely bucolic note to the mechanical sounds of the ship and the cable-laying machinery and the noise of wind and water.

Finally, on July 15, all was ready and the *Great Eastern*—which was carrying, including her own machinery, a total load of 21,000 tons—about what the entire fleet that fought under Nelson at Trafalgar, only sixty years earlier, could have carried—

---

*Indeed, his tombstone would describe him as "the first and greatest of war correspondents."

The Great Eastern *with escorts.*

set sail for Ireland. She soon took the steamer *Caroline,* carrying the heavy shore cable, in tow. At Valentia, a new landing spot for the cable had been selected, Foilhummerum Bay on Valentia Island, where steep cliffs fell to the sea in a scene of wild beauty. Once again there was a large crowd to greet the expedition, with the knight of Kerry again presiding over what looked much like a county fair.

ALTHOUGH A GREAT DEAL HAD BEEN LEARNED about the technology of cable laying, and this fourth attempt was a much more professional affair, amateurs were still very much in evidence. A long line of boats, manned by volunteers, had been arranged like a pontoon bridge to bring the cable ashore. Those onshore wanted to help and clambered into the water to grab the end and haul it in. They didn't realize, however, that enough cable had to be brought ashore to reach up the cliffs and across a couple of fields to the telegraph house. When the end reached the shore, the crowd let out a cheer and the men on the boats, thinking the deed was done, threw the rest of the cable into the water. As a result, the entire operation had to be done again, once the immensely heavy shore cable had been retrieved from the bottom. When all was ready the knight of Kerry gave a speech and the crowd joined him in singing the Doxology. The *Caroline* steamed out to meet the *Great Eastern,* which waited (much to the disappointment of the crowd at Valentia) far offshore, due to its deep draft. When the splice was made, a task that took several hours, the *Great Eastern*, on July 23, 1865, headed for Heart's Content, Newfoundland, as evening approached.

"Happy is the cable laying that has no history," wrote William Russell, and at first all seemed to go well.

> As the sun set a broad stream of golden light was thrown across the smooth billows towards their bows as if to indicate and illumine the path marked out by the hand of Heaven. The brake was eased, and as the *Great Eastern* moved ahead the machinery of the paying-out apparatus began to work, drums rolled, wheels whirled, and out spun the black line of the cable, and dropped in a graceful curve into the sea over the stern wheel. The cable came up with ease from the after tank, and was payed out with the utmost regularity from the apparatus. The system of signals to and from the ship was at once in play between the electricians on board and those at Foilhummerum.[20]

Unlike the early years of the project, there was now a very businesslike atmosphere aboard ship, with everyone either attending to their duties or staying out of the way, thinking only of getting the job done. Even the captain, James Anderson, who had been especially chosen by Field and lent to the expedition by the Cunard Line, wrote later that "I had no mind, no soul, no sleep, that was not tinged with cable."[21]

Only Cyrus Field had no specific duties. The cable laying was officially a project of the Telegraph Construction and Maintenance Company, and Daniel Gooch was on board to make any decisions needed. Field spent part of his time on the bridge with the captain, and part simply walking "Oxford Street," as the crew had dubbed the main deck, two and a half city blocks long, where twin rows of light provided illumination at night. And he spent much time in the test room, where blackout curtains excluded all outside light so that the dancing light from the mirror galvanometer could be seen clearly, telling Thomson and de Sauty that all was well. They had established a schedule for sending test signals to Ireland and receiving them back, once every half hour.

Outside the test room was a brass gong. At the first sign of a faltering signal, someone on duty in the test room was to ring it. Shortly after midnight, when the *Great Eastern* had payed out a mere eighty-four miles of cable, the gong was rung and Captain Anderson ordered the engines to a full stop, although her way carried her a considerable distance farther. The galvanometer indicated that the signal had faltered. It was not a complete break, but a fault in the cable was allowing much of the current to leak out into the sea. Because the resistance of each mile of the cable was known, it was theoretically possible to determine the distance to the fault. But five different people calculated five different distances, ranging from nine to sixty miles.

There was nothing to do but haul up the cable until the fault was located, which was no easy matter. The *Great Eastern* could not maneuver in reverse and, in any event, there was too much chance of fouling the cable in the screw. So the cable had to be cut, and its end laboriously transferred to the bow of the ship, around shrouds, paddle wheels, and lifeboats, a process that took many hours.

When the task was finally accomplished, they were able to retrieve the cable at the rate of only one mile per hour as the machinery, adapted for paying out the cable, was not well designed to haul it in. Instead, a small steam engine, called a donkey engine, was used. Russell wrote that the huge ship daintily picking up the cable "put one in mind of an elephant taking up a straw in its proboscis."[22]

Finally, ten hours and ten miles after the retrieval began, the fault was found. It was ominous. A small piece of wire no larger than a needle had penetrated to the core. Russell, in his notes, called it "flagrant evidence of mischief,"[23] and Daniel Gooch wrote in his diary that the news of this could cause company shares to fall in London. But it could easily have been a happenstance, some momentary carelessness during the eight months it had taken to produce the cable.

The faulty piece of cable was cut off, the end of the cable once more laboriously maneuvered around to the stern and spliced, and the *Great Eastern* once more set off heading west. But the signal again faltered within half a mile. "Even the gentle equanimity and confidence of Mr. Field were shaken in that supreme hour," Russell wrote, "and in his heart he may have sheltered the thought that the dream of his life was indeed but a chimera."[24]

Then, just as suddenly as it had disappeared, the signal returned, as strong as ever. Perhaps, as in 1857, the cold at the bottom of the ocean had sealed whatever fault there was. The *Great Eastern* again increased speed and was soon paying out the cable at a steady six knots. The ship could not have been more suited to the task. "Throughout the voyage," Henry Field wrote, "the behavior of the ship was the admiration of all on board. While her consorts on either side were pitched about at the mercy of the waves, she moved forward with a grave demeanor, as if conscious of her mission, or as if eager to unburden her mighty heart, to throw overboard the great mystery that was coiled up within her, and to cast her burden on the sea."[25] Cyrus Field was particularly grateful for the steadiness that the *Great Eastern* derived from her huge size. For the first time he did not get seasick. But the heavy weather caused the relatively small HMS *Sphinx* to fall behind and be lost to sight. By some unaccountable oversight, she had the only sounding equipment for the expedition. HMS *Terrible* could handle the duty of warning passing ships to steer clear, but the want of sounding gear was to prove costly.

THIS EXPEDITION PRODUCED yet another first for the *Great Eastern*, the first ship's newspaper. Written largely by Russell with drawings by *The Times* of London's artist, Robert Dudley, it was written out in longhand. *The Atlantic Telegraph* printed news and stock prices from Europe received through the cable as well as schedules of the day's activities aboardship, including "From daylight till dusk—Looking out for the *Sphinx*. (Through the kindness and liberality of the Admiralty, this amusement will be open to the public free of charge.)"[26]

All went well for four days, until shortly after noon on Saturday, July 29, when the dreaded gong rang once more and the mighty engines of the *Great Eastern* halted. The line was completely dead. Again the cable had to be cut and hauled around to the bow and retrieved. It was much more difficult this time because there were now over 2,000 fathoms under the ship's keel rather than a few hundred. Four tons of cable were trying to break free. But the crew was also more practiced, and by ten o'clock that night they had reached the fault. The cable was cut again and the end spliced to the cable in the hold.

It was not until Monday that the miles of cable that had been cut, and which were covered in ooze from the sea bottom, were carefully examined. Once again, a wire had been driven into the heart of the cable, shorting it out. All agreed at once that this was clearly sabotage. The wire was exactly the same length as in the previous fault, and the same cable-handling crew had been on duty. Further, sabotage in such a case was not without precedent. A few years earlier, a cable laid across the North Sea to Holland had been sabotaged by having a nail driven through it in order to affect stock prices on the London Exchange.

The crew was summoned and they all, to a man, agreed that it looked like foul play and they further agreed that watches must be set to keep an eye on the men handling the cables.

Again the ship headed for Newfoundland and by Wednesday, August 2, was within six hundred miles of her goal. The *Great Eastern* had paid out twelve hundred miles of cable, and it was working so well that technicians in Valentia were even able to tell when the ship rolled, altering the tiny, whispering effect the earth's magnetic field had on the current in the cable. But that morning, as Cyrus Field was in the cable tier, a grating sound was heard, as of a wire scraping on metal. The crew shouted for men on deck to look out for the cause, but the warning was not heard and the spot passed over the stern and into the water. Immediately a fault was reported in the line. The cable was still functioning, but at a reduced capacity. They could have carried on, but Canning insisted on retrieving the cable.

The matter was considered so minor that the expedition did not bother to inform Valentia that they were going to cut the cable. As this discussion was going on, a workman who had been in the cable tier came up and produced a piece of wire he had found amid the cable that exactly matched the pieces responsible for the two previous faults. It had come from the brittle charcoal iron sheathing. There had been no sabotage after all.

Again the ship slowed to a stop, again the cable was cut and laboriously passed up to the bow and attached to the little donkey engine that would pull it slowly on board. But the donkey engine needed time to get up steam and in the meantime a breeze came up, causing the vast bulk of the *Great Eastern* to drift over the cable, chafing it. When the hauling in began, just as the defective part was coming over the bow, the cable suddenly snapped. It "flew through the stoppers and with one bound leaped over the intervening space and flashed into the sea," Russell reported. "The cable gone! Gone forever down in that fearful depth! There around us lay the placid Atlantic, smiling in the sun, and not a dimple to show where lay so many hopes buried."[27]

A little after noon, most men not engaged in the recovery had gone down to lunch. "Suddenly Mr. Canning appeared in the saloon," Russell wrote, "and in a manner which caused every one to start in his seat, said, 'It is all over! It is gone!' "[28]

The sudden cessation of signals from the *Great Eastern*, news of which reached all over the British Isles within hours, caused a frenzy of speculation as to the cause. Theories ranged from magnetic storms to the sudden foundering of the *Great Eastern* after hitting an iceberg.

On the *Great Eastern* there were really only two choices: abandon the expedition and return to Britain with their tails between their legs, or grapple for the cable and try to fish it out of the sea. Samuel Canning had fished cables out of deep water before and was determined to try again. But this was far deeper water than he had worked in previously. As Henry Field explained, "It was as if an alpine hunter stood on the summit of Mont Blanc and cast a line into the Vale of Chamonix. Yet who shall put bounds on human courage?"[29]

Fortunately, the weather had been brilliantly clear at noon, just before the accident, and Captain Anderson had taken an

Grappling for the cable.

excellent sextant reading of the sun, so they knew where the cable had been lost to within half a mile. They positioned the ship to the south of where the cable lay and a few miles east of where its end was calculated to be.

A length of five miles of wire rope was on board, ready in case it had been necessary to buoy the cable in a bad storm. To this was attached a grappling hook, which looked a bit like an anchor, but with five arms, each tapering to an end shaped like a shovel. Russell described them as "the hooks with which the Giant Despair was going to fish from the *Great Eastern* for a take worth, with all its belongings, more than a million."[30]

Thrown overboard about three in the afternoon, it took fully two hours for the grapple to reach the bottom. Because there was no sounding equipment on board, the depth could only be guessed at. With the grapple on the bottom, the ship began sailing northward, searching for the cable. Near morning, the rope quivered and the head of *Great Eastern* was tugged slightly around. They had hooked something, but was it the cable? The answer was soon clearly yes. Had it been a piece of debris or an old wreck, the weight would have diminished when the wire rope shortened as they hauled it up. But the strain steadily increased as the grapple rose and more and more cable was lifted off the bottom. Also, the contact had been made exactly where Captain Anderson's sighting indicated the cable lay.

The wire rope, unfortunately, was not continuous. Instead, it was made of hundred-fathom lengths shackled together. The shackles proved to be the weak points. One broke but the brakeman was able to clamp it before the end vanished into the sea, but not before two workmen were severely lacerated by the end, which acted like a steel bullwhip. Then, when they had raised the cable about three-quarters of a mile from the bottom, another shackle broke and the cable plunged back into the depths, taking two miles of rope with it. A second attempt succeeded in hooking the cable, but this time a shackle broke when it had raised it a mile. In a third attempt, a fluke caught on the rope, a fact discovered only when the rope was raised after failing to catch the cable. A final attempt was made on Friday, August 13. Again, after pulling in about eight hundred fathoms, a shackle gave way.

There was now not enough rope left to make a fifth attempt. There was nothing to do but return once more to Britain in failure.

As the *Great Eastern*'s engines began to pound, the *Terrible* signaled, "Farewell!"

"Good-by. Thank you," the melancholy ship signaled back.

"THE *GREAT EASTERN* LOOMS ALL GLORIOUS IN THE MORNING SKY"

CYRUS FIELD EXPECTED to return to the British Isles once more the object of derision. He was, therefore, pleasantly surprised to find himself instead almost a hero. The New World had had no news of the latest expedition for many days and did not even know when it might learn what was happening. Indeed, not until HMS *Terrible* reached St. John's, Newfoundland, did it learn both that the expedition had sailed from Ireland and of the latest failure.

But the Old World, connected to the *Great Eastern* by a telegraph cable that worked to perfection, had followed the progress of the expedition with keen interest and increasing expectations of success. Then, suddenly, on the morning of August 2, silence, and the silence continued for two long weeks as the *Great Eastern* fought to wrestle the cable back to the surface and only then sailed sadly for home.

In the absence of any solid information, speculation, of course, ran rampant. The *Great Eastern*, too big to be structurally sound, had hogged on an Atlantic swell, broken her back, and plunged to the bottom with all hands. She had struck an iceberg in the Labrador Current and similarly foundered. Only on the sixteenth day of silence, as the *Great Eastern* approached the small Irish port of Crookhaven, at the southwest tip of the island, was the truth learned, when an enterprising local journalist sailed out to meet them. "We thought you went down,"[1] he yelled when he got within hailing distance. When he came aboard, he heard the whole dramatic story and hastened back to Crookhaven to relay it by telegraph, while the *Great Eastern* sailed on to the Nore in the Thames Estuary.

Besides the relief of learning that five hundred men—including some of Britain's most distinguished—and her greatest ship had not vanished without a trace, it was immediately clear that success had been fully within reach before simple bad luck intervened. Field and others had spent the time between the final failure and reaching England drawing up a new prospectus. It made twelve points, including,

- That the 1858 cable had proved that it was possible to lay a cable across the Atlantic and to send telegraph signals through it.
- That the *Great Eastern*, "from her size and constant steadiness, and from the control over her afforded by the joint use of paddles and screw, renders it safe to lay an Atlantic Cable in any weather."[2]
- That the cable, if lost, can be located and, with proper equipment, raised to the surface.

"All who were on that voyage," Henry Field wrote, "felt a confidence such as they had never felt before."[3] Field was delighted to discover that the company directors in London were of the same opinion. Indeed, they quickly decided that success the next year was so certain and the business potentially so large, that they should lay a whole new cable the next year and then return to midocean, raise the cable lost in 1865, and complete it.

The Telegraph Construction and Maintenance Company

again offered to lay the new line, estimated to cost £500,000, and take their profit in Atlantic Telegraph stock at the rate of 20 percent of the cost of laying the line.

The Atlantic Telegraph Company decided to raise £600,000 by issuing 120,000 shares of preferred stock, to pay the junk-bond rate of 12 percent, at £5 per share. Everything seemed on track for a successful expedition the next year and Field returned to the United States in September. A fellow passenger wrote, "We felt the deepest sympathy for him, and to our surprise he was the life of the ship and the most cheerful one on board."

Field explained his cheerfulness by telling him, "We've learned a great deal, and next summer we'll lay the cable without a doubt."[4] Whether his fellow passengers, certainly aware of the many failures, believed him is a good question.

Field was returning to New York because he wanted to be home for his daughter Isabella's marriage in October. But there was little for him to do stateside other than to meet with his fellow directors of the New York, Newfoundland, and London Telegraph Company. The directors agreed that the cable laid across the Cabot Strait back in 1855 should be replaced. It was showing signs of faltering, and cable technology had advanced so much in the decade since it had been laid. They decided to order ninety miles of new cable from the Telegraph Construction and Maintenance, to the same specifications as the new Atlantic cable.

Ever restless, however, Field was soon back in England, arriving on Christmas Eve. There he found that a new problem had befallen the Atlantic Telegraph Company.

This time it was not technological or financial but legal. Two days earlier the attorney general had ruled that the company, which had been created by private act of Parliament, as many companies still were at that time, rather than by registration under the Companies Act of 1844, had no right to issue new stock without further Parliamentary action. That was currently impossible, as Parliament was in recess, and in any event the deadline for submitting private bills in that session had passed.

In this case, there was nothing to do but stop everything and return the money that had already been subscribed until the lawyers found a solution. Daniel Gooch, the chairman of the Great Western Railway and the head of the syndicate owning the *Great Eastern*, suggested simply establishing a new company, this one under the Companies Act, in order to get the job done, and straighten out the legal tangles later. It was to be capitalized at £600,000 and Gooch offered to subscribe £20,000 himself. Field subscribed £10,000. The new company was named the Anglo-American Telegraph Company. Because the directorship of that company, as well as those of the Atlantic Telegraph Company, the New York, Newfoundland, and London Telegraph Company, and the Telegraph Construction and Maintenance Company, overlapped to a great extent, they operated effectively as one entity.

Thomas Brassey quickly agreed to invest up to £60,000 in the new company and various directors of the other companies agreed to put up £10,000 each. So by the time the subscription was opened to the public, there was a total of £230,500 already in the till, and the whole amount needed was raised in a couple of weeks. Given the fact that the British economy was in recession in the mid-1860s, the ease with which the money was raised this time clearly indicated that there was now a wide perception that this was a proven technology. It was just a question of getting the job done.

A NEW CABLE WAS ORDERED and the Telegraph Construction and Maintenance Company began turning it out at the rate of

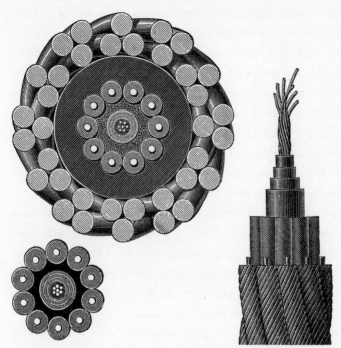

The heavy shore end (top) and main ocean cable used in 1865 and 1866.

twenty miles per day. Although it had the same dimensions as the cable from the previous year, it was better designed owing to the experience gained in laying that one. The iron armoring was now galvanized, the wires being passed through molten zinc to produce a coating of zinc-iron alloy that made them impervious to rust and corrosion. It also made them more ductile—able to stretch further without breaking. The new cable could take nearly half a ton more strain before breaking.

Because the armor was now rustproof, the hemp did not have to be impregnated with pitch. This meant that the new cable was rope colored, not black, and was not sticky. Stray bits of iron wire were far less likely to adhere to it and cause mischief. It also weighed more than four hundred pounds less per mile.

The paying-out machinery was given more power so that it could haul in, if necessary, as well as pay out, and the tedious, risky business of shifting the cable around to the bow was obviated. And the *Great Eastern* was now supplied with adequate wire rope for hauling cables off the sea floor. Willoughby Smith, the head electrician with the Telegraph Construction and Maintenance Company, devised a means of testing the insulation of the cable continuously so that a fault would show up instantly, rather than after as much as half an hour. The equipment Smith developed was so sensitive that it would detect faults, "even when they are so small that they would not weaken the signals through the Atlantic cable one millionth part!"[5]

The *Great Eastern* herself was also altered. The stern was fitted with a "crinoline" to ensure that the cable would not foul the screw, and the paddle wheels were uncoupled, making it possible to use just one at a time, thus greatly increasing the ability of the ship to steer while steaming in reverse. The great ship also had her bottom scraped, removing tons of marine growth and doubtless considerably increasing her top speed, not that that was needed for a cable-laying operation. This was, however, no easy task as the ship could be neither dry-docked nor careened. But Captain Anderson devised a "simple instrument" by which she could be raked and scraped from the deck.

Taking no chances, the cable crews in the next expedition were to wear canvas suits that fastened up the back and had no pockets, making it nearly impossible to bring instruments for sabotage in close proximity to the cable.

The technicians were to be the same as the previous year, except that Willoughby Smith was the official electrician, "advised" by William Thomson. Field again was the only American on board.

Field's presence was well recognized as key to the venture's success. After Field returned from a trip to New York in May, William Thomson wrote him that "I am very glad to learn that you are again in this country. You are not come too soon as the A.T.C. [Atlantic Telegraph Company] seems to require an impulse, and I am sure will be much the better for your presence."[6] Earlier, Captain Anderson had written after Field's return from an earlier trip, "I feel as if our watch had got its mainspring replaced, and had been trying to go without it for the last three months."[7]

The *Great Eastern* sailed from the Thames Estuary on June 30, accompanied by the chartered vessels *Albany* and *Medway*. She was so deeply laden that she barely cleared the sandbar at the Nore. The little lightship at Margate managed to produce an impromptu band that serenaded the *Great Eastern* with "Goodbye, Sweetheart, Good-bye!" as she steamed majestically to sea, her great paddle wheels thrashing through the water.

On July 4, off Fastnet Rock on the southern coast of Ireland, "the American Flag was hoisted and speeches made in compliment of Mr. Field."[8] Then, with the Victorian love of amateur theatrics, a makeshift theater called the Great Atlantic Haul was organized and a satiric review presented called "A Cableistic and Eastern Extravaganza." Field was the butt of most of the jokes. Referring to the cable stowed below, the character of Field was made to say

> Oh! That I could shuffle off this mortal coil!
> (*À la Hamlet*) I've furrowed the Atlantic many times,
> And 'mid such toil have held convivial dinners.
> For me the sparkling wine cup nightly flowed,

Robert Dudley's painting of laying the shore end at Foilhummerum Bay.

And often flowed in vain;
While others to the joy of music clung,
I plied the bottle, but 'twas then of ink.
Prospectuses I drew; percentage showed,
And e'en through worse times on 'change
Could lead grave, bearded men
To wander forth and muse
On the triumphant joys I promised them.9

On Friday, July 13, the shore cable was once again hauled ashore at Foilhummerum Bay and the cliffs there were alive with onlookers. Henry Field, putting on his clergyman's hat, held services, praying for final success. The *Great Eastern* remained at sea, thirty miles out. The weather was foggy and it took some time to find the buoy that marked the seaward end of the shore cable, raise the cable, and get it spliced. The splicing itself was no small matter. As Henry Field described it, "Quick, nimble hands tore off the covering from some yards of the shore end of the main cable, till they came to the core; then, swiftly unwinding the copper wires, they laid them together, twining them closely and carefully as a silken braid. Thus stripped and bare this new-born child of the sea was wrapped in swaddling-clothes, covered up with many coatings of gutta-percha, and hempen rope, and strong iron wires, the whole bound round and round with heavy bands, and the splicing was complete."[10]

The line was tested through its whole length and found to be in perfect working order. Then HMS *Terrible* led the way, followed by the *Great Eastern*, which was flanked by the *Albany* and *Medway*.

The track of the 1866 expedition was some thirty miles south of the 1865 cable to ensure that when they later grappled for

The splice is made linking the Valentia end and the ocean length of cable.

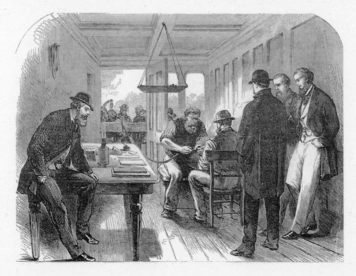

the old cable they did not accidentally foul the new one. The *Great Eastern* was able to maintain six knots easily with only her paddle wheels, and her screw was disengaged. The fog and rain of Friday, July 13 (the superstitious consoled themselves with the knowledge that Columbus had also sailed on a Friday), soon cleared and the weather became perfect. "The sea like a mill-pond,"[11] wrote John C. Deane, secretary of the Anglo-American Telegraph Company in his diary. William Howard Russell was not aboard this trip, because *The Times* of London had sent him to cover the Seven Weeks' War between Prussia and Austria, and diaries of Deane and Gooch are the sources of most details on the 1866 expedition.*

On Sunday a man fell overboard from the *Terrible* but was quickly rescued. Monday was perfect. Indeed, Deane wrote in his diary that "it was so calm that the masts of our convoy were reflected in the ocean, an unusual thing to see. A large shoal of porpoises gamboled about us for half an hour. A glorious sunset, and later, a crescent moon, which we hope to see in the bright-ness of her full, lighting our way into Trinity Bay before the days of this July shall have ended."[12]

Early Wednesday morning, however, the cable got tangled in the after tank when a "foul flake" occurred. Fortunately, the ship's engines were stopped at once and then reversed to make

---

*The Prussians had paid close attention to the American Civil War, sending many military observers. They thus had learned firsthand the military uses of the telegraph and the railroad. Field Marshall Von Moltke was able to concentrate his forces quickly by railroad, coordinating them from Berlin by telegraph. The result was the Battle of Königgrätz, in which Austria's army was decisively defeated and Austria was forced to seek terms. Ironically, a breakdown in the telegraph nearly caused the Prussians to lose the battle when one army, for lack of orders, failed to move up. Only a message, delivered the old-fashioned way, by a dispatch rider who galloped twenty miles, saved the day.

sure that not too much strain was placed on the cable. As Captain Anderson expertly kept the immense ship stationary by slight adjustments of paddle wheels and screw, Samuel Canning calmly directed the cable crew in disentangling the mess. "During all this critical time," Deane wrote in his diary, "there was an entire absence of noise and confusion. Every order was silently obeyed, and the cable men and crew worked with hearty good-will. Mr. Canning has had experience of foul flakes before, and showed that he knew what to do in the emergency."[13]

The grotesquely amateur attempt to lay the cable across the Cabot Strait in 1855 — while women in full skirts strolled the deck, Newfoundland dogs galumphed about, and Samuel Canning had been grateful not to have been sacked on the spot — had given way to cool competence and great professional skill.

On Saturday, July 21, the *Great Eastern* passed the halfway point, where the *Niagara* and *Agamemnon* had twice seen the 1858 cable part. There was a growing sense of optimism among the technical staff and the crew alike, but the tension never abated. "I now hear the rumble of the cable over my head in the cabin," Gooch wrote in his diary once the cable from the after tank was exhausted and the forward tank was being used, "and am constantly listening to it. This stretch of the nerves day after day is hard work, and the mind has no change; morning, noon, and night it is all the same — cable, cable, cable."[14]

Field, as always, was indomitably optimistic. On Monday, July 23, he telegraphed to Valentia to get "the latest news from Egypt, China, India, and distant places for us to forward to the United States on our arrival at Heart's Content."[15] Perhaps anticipating one of the major demands for the Atlantic cable, he also asked for the latest financial news, money rates, and cotton prices. By Tuesday he estimated reaching Newfoundland on Friday and asked William

Glass, in Valentia, when he thought the cable might be opened for public business.

Glass's reply could only have been music to Field's ears. "If you land the cable on Friday, I see no reason why it should not be open on Saturday."[16]

On Tuesday they passed the point where the 1865 cable had parted, and still everything went well. "May we have three days more of such delightful monotony!" Deane wrote in his diary that day.[17]

As they neared Newfoundland the weather deteriorated into rain and dense fog.* The telegraph fleet rang with signal whistles—two for the *Terrible*, three for the *Medway*, four for the *Albany*, and one, by far the loudest, for the *Great Eastern*—to ensure that they did not collide.

The admiralty had ordered Sir James Hope, admiral on the North American station, to place a ship thirty miles off the entrance to Trinity Bay to help guide the fleet in. On Thursday, July 26, the *Albany* was dispatched to look for her.

The next day, the fleet spotted two ships bearing down on it. It was the *Albany* and HMS *Niger*, of the North American station. The fog hung thick all night as the *Great Eastern* crept toward the shore. Then, early the next morning, Deane noted that "good fortune follows us, and scarcely has eight o'clock arrived when the massive curtain of fog raises itself gradually from both shores of Trinity Bay, disclosing to us the entrance to Heart's Content."[18]

The tiny hamlet had been dressed with flags for their arrival. Because there had been no news since several days before the *Great Eastern* left Ireland, the reporters in Heart's Content had had

---

*The meeting of the warm Gulf Stream and the cold Labrador Current makes the Grand Banks off Newfoundland among the richest fishing grounds on earth and also one of the foggiest.

The Great Eastern *at Heart's Content.*

nothing to do but wait. For "the correspondents of the American papers," Henry Field reported,

> only a long and anxious suspense, till the morning when the first ship was seen in the offing. As they looked toward her, she came nearer—and see, there is another and another! And now the hull of the *Great Eastern* loomed up all glorious in the morning sky. They were coming! Instantly all was wild excitement on shore. Boats put off to row towards the fleet. The *Albany* was the first to round the point and enter the bay. The *Terrible* was close behind. The *Medway* stopped an hour or two to join on the heavy shore end, while the *Great Eastern*, gliding calmly in as if she had done nothing remarkable, dropped her anchor in front of the telegraph house, having trailed behind her a chain of two thousand miles, to bind the Old World to the New.[19]

If the iron soul of the *Great Eastern* thought she had done nothing remarkable, *The Times* of London disagreed. "It is a great work," *The Times* stated, "a glory to our age and nation, and the men who have achieved it deserve to be honored among the benefactors of their race."[20] The news of the success of the cable expedition received a more prominent place in the paper than the headline "Treaty of peace signed between Prussia and Austria!"

Queen Victoria also thought the accomplishment memorable and showered those involved with honors. William Thomson, Samuel Canning, William Glass, and Captain Anderson all received knighthoods. Daniel Gooch and Curtis Lampson, deputy chairman of the Atlantic Telegraph Company, were created baronets. Again it was noted that Cyrus Field was not honored by the queen only because he was not a British subject, but the British newspapers quickly dubbed him "Lord Cable," and Congress voted him a gold medal (although, because of a bureaucratic mix-up, it was several years before he actually received it). Constantino Brumidi, who painted the murals in the United States Capitol, depicted Venus holding the Atlantic cable on the interior of the dome, where it can still be seen from the Rotunda.

Heart's Content was as frenzied as a fishing village of some

A gold medal awarded to Cyrus Field by Congress in March 1867.

sixty houses and a church on the shore of a wilderness could get. "There was the wildest excitement I have ever witnessed," Daniel Gooch wrote in his diary. "All seemed mad with joy, jumping into the water and shouting as though they wished the sound to be heard in Washington."[21]

PREPARATIONS WERE MADE TO SPLICE THE CABLE that reached across the Atlantic onto the shore cable already laid, and Cyrus Field had gone ashore. There he learned that the cable across the Cabot Strait had broken down, "so that a message which came from Ireland in a moment of time," his brother reported, "was delayed twenty-four hours on its way to New York."[22] Field had, of course, anticipated this possibility and had urged the New York, Newfoundland, and London Telegraph Company to keep the original cable in repair so that it would serve until the new one ordered in England could be laid. Instead, the company had decided to wait and see how the Atlantic cable expedition fared.

Field immediately telegraphed to St. John's to charter the steamer *Bloodhound* to fish up the old cable and get it working again. In the meantime, he also chartered the *Dauntless* to run back and forth between Newfoundland and Nova Scotia carrying messages the old-fashioned way. It was quickly determined that the old cable had been fouled by an anchor a few miles from shore and broken. A piece of the new cable was spliced onto both ends and service was quickly restored.

Because of the broken cable, it was not until Sunday, July 29, that the news reached New York, when Field's message to the Associated Press arrived. "Heart's Content, July 27," it read. "We arrived here at nine o'clock this morning. All well. Thank God, the cable is laid, and is in perfect working order."23

With the cable across the Cabot Strait repaired quickly, messages of congratulations began to pour into Heart's Content, which was, momentarily, the center of the world. Queen Victoria and President Andrew Johnson exchanged congratulations. Perhaps some measure of what he had managed to accomplish came home to Cyrus Field when he received, nearly simultaneously, two messages, one from San Francisco and another from Ferdinand de Lesseps, builder of the Suez Canal, who was in Alexandria, Egypt, each thousands of miles away from Heart's Content, Newfoundland.

Commercial traffic also immediately began to utilize the Atlantic cable. Daniel Gooch, now a major investor in the Anglo-American Telegraph Company as well as head of the syndicate that owned the *Great Eastern*, was delighted to note in his

diary that "yesterday we had fifty messages, paying us, I suppose not less than twelve thousand pounds."[24]

While Daniel Gooch was happily counting his profits, the world was already beginning to change. As early as August 16, 1866, only two weeks after the cable had opened for business, *Harper's Weekly* reported that "the monetary quotations in this city and London are becoming equalized."[25]

A market can never be larger than the area in which communication is effectively instant. That meant that until the mid-nineteenth century, every major city had its own financial market. In the 1840s, the rapid spread of the telegraph on land had allowed the largest markets, such as London and Wall Street, to become dominant. Now, thanks to the cable, these two markets could begin to act together.

IMMEDIATELY UPON SUCCESS, Field had telegraphed to his wife, who had stoically endured the many separations from her husband, that "now we shall be a united family." But the very next sentence belied that hopeful statement. "We leave in about a week to recover the cable of last year."[26]

The *Great Eastern*, its fuel capacity greatly reduced by the vast cargo of cable, had to take on eight thousand tons of coal sent over from England by collier before proceeding. Also, the six hundred miles left of the 1865 cable that had been stored in the *Medway* was now transferred to the *Great Eastern* to be spliced on to the cable when it was raised from the ocean bed. She departed Heart's Content on Thursday, August 9.

The *Albany* and the *Terrible* had departed eight days earlier to search for the lost cable. They quickly found it, and succeeded in

Banquet at the Metropolitan Hotel in honor of Cyrus Field.

raising it a few hundred fathoms. But bad weather moved in and they had to buoy it. The weather then tore away the buoy and the cable sank once more to the bottom, taking two miles of wire rope with it. When the *Great Eastern* joined them on Sunday, August 12, everyone expected that the cable would be quickly located again. It was not. Despite repeated drags of the grapnel ("I often went to the bow," Field reported, "and sat on the rope, and could tell by the quiver that the grapnel was dragging on the bottom two miles under us"[27]), the cable was only hooked on Thursday. It took fourteen hours to reel her in and it finally appeared on the morning of August 20. "We had it in full sight for five minutes," Field wrote, "a long, slimy monster, fresh from the ooze of the ocean's bed—but our men began to cheer so wildly that it seemed to be frightened, and suddenly broke away and went down into the sea."[28]

Again and again the expedition hooked the cable only to have it break before reaching the surface. Evidently the cable was

The final tactics adopted for picking up the 1865 cable.

weaker after its yearlong immersion and could not bear its own weight in such a deep part.

Finally, another approach was tried. The fleet moved ninety miles east, where the water was shallower by six hundred fathoms. They quickly hooked the cable (it was the thirtieth time the grapnel had been lowered), but this time brought it only halfway to the surface and then buoyed it, still nine hundred fathoms below the surface. The *Great Eastern* then moved several miles farther west, toward the cable's lost end, and hooked the cable again. The *Medway*, stationed two miles farther west still, also hooked the cable. The two ships began to haul it in very carefully. When the *Medway* had brought her to within three hundred fathoms of the surface, while the *Great Eastern*'s grapnel was still eight hundred fathoms down, the *Medway* deliberately broke the cable to lighten the load. The eighty-odd miles of cable from this position to where it had broken the previous year were abandoned.

The *Great Eastern* began to haul the monster in, and finally, at one o'clock on the morning of Sunday, September 2, the end came aboard and was quickly secured. The cable was hooked up to the testing room and a message was sent to Valentia. Although Valentia had been expecting it, the signal was still thrilling when it came after a year of silence.

The London *Spectator* reported from Valentia,

Night and day, for a whole year, an electrician has always been on duty, watching the tiny ray of light through which signals are given, and twice every day the whole length of wire—one thousand two hundred and fifty miles—has been tested for conductivity and insulation. . . . Sometimes, . . . wild incoherent messages from the deep did come, but these were merely the results of magnetic storms and earth-currents, which deflected the galvanometer rapidly, and spelt the most extraordinary words and sometimes even sentences of nonsense. Suddenly, one Sunday morning, while the light was being watched by Mr. May, he observed a peculiar indication about it, which showed at once to his experienced eye that a message was at hand. In a few minutes the unsteady flickering was changed to coherency . . . and the cable began to speak the appointed signals, . . . "Canning to Glass" must have seemed like the first rational words uttered by a high-fevered patient, when the ravings have ceased and his consciousness returns.[29]

On board the *Great Eastern,* jubilation erupted when contact with Ireland was made. Robert Dudley, the artist sent by *The London Illustrated News,* reported that "along the deck outside, over the ship, throughout the ship, the pent-up enthusiasm overflowed; and even before the test-room was cleared, the roaring bravos of our guns drowned the huzzahs of the crew, and a whiz of rockets was heard rushing high into the clear morning sky to greet our consort-ships with the glad intelligence."[30]

For Cyrus Field, it was, perhaps, the supreme moment of his life.

Never shall I forget that eventful moment, when, in answer to our question to Valentia, whether the cable of 1866, which we had a few weeks previously laid, was in good working order, and the cable across the Gulf

The Eighth Wonder of the World.

of St. Lawrence had been repaired, in an instant came back those six memorable letters, "Both O.K." I left the room, I went to my cabin, I locked the door; I could no longer restrain my tears — crying like a child, and full of gratitude to God that I had been permitted to witness the recovery of the cable we had lost from the *Great Eastern* just thirteen months previous.[31]

The expedition laid the last of the 1865 cable without further trouble. Thus the second telegraph cable to cross the Atlantic Ocean was soon in place and working only four weeks after the first cable had begun to operate. Besides doubling the capacity, the second cable was a valuable insurance policy, allowing continuous operation of the system even if one cable was down for repair or maintenance.

As FIELD PREPARED to depart the *Great Eastern* for the last time, Captain Anderson cried out, "Give him three cheers!" The little harbor of Heart's Content rang with them. "And now three more for his family!" and again the pine-covered hills echoed the heart-felt sentiments of all who were present. Then "the wheels of the *Great Eastern* began to move, and the noble ship, with her noble company, bore away for England."[32]

As *The Times* of London had said sixteen years earlier, when the first submarine cable had been laid across the English Channel, "The jest of yesterday has become the fact of today."[33] Never again would North America be out of instant communication with Europe for more than a few hours.

EPILOGUE

"THERE WERE TWO WORLDS . . . LET THERE BE ONE"

IT WAS NOT IN THE NATURE of Cyrus Field to rest on his triumph, but his days as an entrepreneur were largely over. World-famous and greatly admired on both sides of the Atlantic, he lent his name and efforts to many worthy causes, including the settling of the dispute between Britain and the United States regarding the *Alabama*.

In business, he was largely content to invest in going concerns. The cable made him seriously rich once again and he was able to pay off his debts in short order. Soon he built a large country house on the Hudson River north of New York, naming it Ardsley after the town where John Field, the astronomer, had been born in England.

But, like many entrepreneurs, Field was not a good investor. In the 1880s he was a major player on Wall Street, interested in such companies as the Manhattan Elevated Railway Company. Indulging increasingly in speculation, often with his sons, Cyrus Jr. and Edward, he was nearly ruined when the market turned against him. On a single, ghastly day, June 24, 1887, he lost about $6 million.

His last days were ones of sadness. A daughter was severely mentally ill and his favorite son, Edward, was found to have been engaged in stock fraud. His beloved wife, Mary, died shortly after their fiftieth wedding anniversary, and Cyrus followed her the next year.

Although his fortune had disappeared in the cutthroat Wall Street of the late nineteenth century, his fame, of course, endured. When he died, on July 12, 1892, age seventy-two, a service was held at Ardsley attended by hundreds, including the famous as well as the ordinary. His body was then taken to Stockbridge, Massachusetts, where he had been born into a very different world, and after a brief service in the church where his father had preached, he was laid to rest beneath a tombstone that reads

Cyrus West Field
To whose courage, energy and perseverance
the world owes the Atlantic telegraph

It is a simple monument, but all that is necessary. For as a contemporary said, "Every message that flashes through the Atlantic cable is his eulogy." And that eulogy continues unabated, and, indeed, far greater than Field could ever have imagined.

ALTHOUGH THE SUCCESSFUL COMPLETION of two cables across the Atlantic Ocean in 1866 proved the technical feasibility of long-distance submarine telegraphy, its commercial viability was not yet assured. There is always a shakedown period in any new technology while the best ways to maximize profit are explored. This was quite as true of the Atlantic cable in the 1860s as it would be of the Internet in the 1990s.

At first, the rates were set extremely high—$10 a word with a ten-word minimum. As $10 was a good weekly wage for a skilled workman at that time, these rates eliminated all but large businesses and the very rich from using the cable. Still, between July 28 and October 31, 1867, the Atlantic cable transmitted 2,772 commercial messages across the Atlantic, and this produced enough revenue to be profitable, averaging $2,500 a day. At this rate, however, only about 5 percent of the cable's capacity was being utilized. So the company cut the rate in half, charging $46.80 for a ten-word message, and revenue increased to an average of $2,800 a day as many more people began to utilize the cable.

It was only when competition reared its head that prices began to fall rapidly and usage exploded. In 1869 a new company, the Société du Câble Trans-Atlantique Français, laid a cable from Brest, France, to the French island of St. Pierre, south of Newfoundland, and then on to Massachusetts. The Anglo-American Telegraph company, naturally, opposed this new enterprise, which was largely funded by British capital, but had no means to prevent it. The *Great Eastern* was once again employed to lay it.

Prices continued to fall as more cables were laid and ways were developed to lower costs. In 1870, for instance, Horace Greeley of the *New York Tribune* spent $5,000 to transmit a single dispatch regarding the Franco-Prussian War. The demand of the public for up-to-the-minute news from Europe forced the ferociously competitive major newspapers to incur such expenses. But they quickly agreed to form United Press International in order to share the costs.

By 1870, Wall Street, where knowledge of the latest London prices could mean the difference between fortune and bankruptcy, was spending a million dollars per year on cable charges and London brokers were spending similar sums. The ever-growing demand caused more and more cables to be laid in different areas of the world.

The *Great Eastern* laid a total of five cables across the Atlantic in these years, and by 1900 there would be a total of fifteen, including a cable to Brazil and Argentina. And the Atlantic was not the only ocean beneath which a cat's cradle of submarine cables was being laid. In 1869 Sir John Pender, an investor in the Atlantic cable, organized the British Indian Submarine Telegraph Company and laid a cable from Suez to Bombay. This finally gave Britain a secure telegraphic link to its Indian empire. The original British telegraphic link with India had been a submarine cable from India only to the Persian Gulf and from there went over land through the Turkish Empire and a number of European countries. It often took a week for a message to travel the whole route and was frequently garbled as it was relayed over and over by clerks who did not speak English.

Pender did not stop with India. By 1871 Australia was reached, via Singapore, and lines soon stretched to China and Japan. In 1902 the globe was at last girdled with submarine cables when a line was run from Vancouver, British Columbia, to Australia and New Zealand.

Thus, within a generation, submarine telegraphy had made it possible for every major nation in the world to be in nearly instant communication with every other. These cables had not only become vital parts of the modern world—indeed, helped powerfully to bring the modern world into being—they had also become important national assets (and targets) in time of war. Within hours of Britain's declaration of war against Germany, in August 1914, a British cable-laying ship, the *Teleconia,* was off the coast of Germany in the North Sea. Methodically her crew grappled for and raised the five German transatlantic cables. As each was raised, it was cut and allowed to slide back into the sea.

At a stroke, Germany was forced to use the new wireless technology developed two decades earlier by Guglielmo Marconi in order to communicate with the outside world, and British intelligence could intercept these messages and decode them. In 1917, one of these messages proved fatal to Germany. The "Zimmerman Telegram," which offered Mexico an alliance and promised the return of her "lost provinces" from the United States, finally persuaded President Woodrow Wilson to ask Congress to declare war.

WIRELESS TELEGRAPHY, the first practical use of the transient electrical currents that William Thomson had investigated in the 1850s, quickly became an alternative technology to submarine cables, but never replaced them, even for nonsecret communications. Subject to storms and magnetic disturbances, wireless could not provide anywhere near the same reliability as cables. Even with the advent of satellites, transoceanic communication is still largely by cable. Satellites, subject to solar "weather," account for only about 30 percent of global communications.

The Atlantic cable system as of 1903.

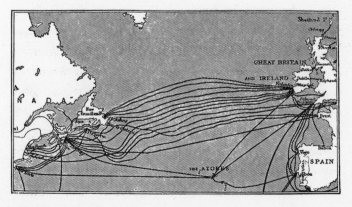

Only in the dawning technology of telephony did noncable means compete with cable, when radiotelephony—utilizing radio to transmit telephone calls—was developed in the 1920s. Transatlantic telephone calls first became possible by this method in 1926, and cost a princely $75, in 1926 dollars, for three minutes.

The reason is that telephone signals must be much stronger than telegraphic ones in order to be reliably converted back to a voice at the other end. There was then no way to transmit a signal of sufficient strength through a transoceanic cable. The first telephone cable, laid in 1921, ran only the ninety miles from Havana to Key West, and allowed only three calls simultaneously.

However, there are strict limits to the number of radio frequencies that can be utilized by radiotelephony, and increased demand for transoceanic communications quickly reached those limits. The problem of sending telephone signals through submarine cables was finally solved when AT&T developed the "regenerator," a device that allowed the signal to be renewed periodically on its journey through the cable. The first transatlantic telephone cable was laid in 1956, capable of transmitting thirty-two calls simultaneously. With the advent of the electronics revolution, cable capacity grew by leaps and bounds. By 1988 cables were routinely carrying ten thousand simultaneous calls, and by 1996, a new trans-Pacific cable could carry 320,000.

The increased capacity led to sharply lower prices, resulting in much higher usage. In 1950 a million overseas phone calls originated in the United States. By 1970 the number had risen to 23.4 million. In 1980 it was 200 million. In 1997 a staggering 4.2 *billion* overseas phone calls originated in the United States. The numbers continue to surge as data transmissions and Internet usage add to demand. A new fiber-optics cable network between the United States and China, now in the planning stages, will cost one billion dollars. Allowing for inflation, that is, very, very roughly, ten times what the first transatlantic cable cost to build and install 136 years ago. But its capacity is billions of times as great.

It will have a total length of 18,750 miles and will have a capacity of 80 gigabytes per second, enough to carry 4 million simultaneous phone calls or transmit data equal to about 18,000 encyclopedia volumes per second. Similar cables are planned to run from the United States to Brazil and from the United States to Europe.

Today it costs less to make a phone call from New York to London than it did to call from New York to Philadelphia half a century ago.

UPON FIELD'S SUCCESS IN 1866, the distinguished New York lawyer William Maxwell Evarts (whose father had once roomed with Field's father at Yale) toasted him at one of the inevitable banquets. Evarts, who would be named secretary of state in the Grant administration, was already deeply concerned with international affairs. He noted that on the statue of Christopher Columbus in Genoa, Italy, is the inscription "There was one world; he said, 'let there be two,' and there were two." "Let us now, then," said Evarts, "say of the Atlantic Cable and its author: 'There were two worlds; *he* said, 'let there be one again,' and they were one."

This is far more true today. Cyrus Field, like all entrepreneurs, sought only to succeed in his particular endeavor, laying a telegraph line across the Atlantic Ocean, and to create a profitable enterprise by doing so. But because Adam Smith's "invisible hand" is ever at work, he did much more than gain a fortune for himself. He laid down the technological foundation of what would become, in little over a century, a global village.

NOTES

⊶ I ⊷
"An Enterprise Worthy of This Day of Great Things"

1. A photocopy of the deed is in the possession of the author. Thomas Nightingale happens to be the present author's fifth great-grandfather.

2. Quoted in Houghton, p. 183.

3. The phrase is in the novel *2001: A Space Odyssey*.

4. Quoted in Bright, p. 19.

5. Quoted in Carter, p. 209.

⊶ II ⊷
Cyrus Field

1. Quoted in Judson, p. 2.

2. Quoted in Carter, p. 22.

3. Quoted in Carter, p. 23.

4. Quoted in Judson, p. 3.

5. Quoted in Carter, p. 30.

6. Quoted in Carter, p. 23.

7. Quoted in Gordon, p. 30

8. *Harper's Monthly,* June, 1856.

9. Quoted in Judson, p. 19.

10. Quoted in Judson, p. 23.

11. Quoted in Judson, p. 25.

12. Quoted in Voorsanger and Howat, p. 313.

13. Judson, p. 36.

14. Quoted in Carter, p. 79.

15. Judson, p. 56.

→ III ←
Newfoundland

1. Quoted in Field, pp. 7–8.
2. Quoted in Field, p. 10.
3. Field, p. 17.
4. Quoted in Carter, p. 99.
5. Quoted in Field, pp. 19–20. Italics in original.
6. Quoted in Field, p. 21.
7. Quoted in Judson, p. 62.
8. Quoted in Field, p. 28.
9. Quoted in Carter, p. 101.
10. Quoted in Field, pp. 32–33.
11. Quoted in Carter, p. 102.
12. Quoted in Carter, p. 104.

→ IV ←

"How Many Months? Let's Say How Many *Years!*"

1. Field, p. 41.
2. Field, p. 42.
3. Quoted in Carter, p. 111.
4. Quoted in Carter, p. 116.
5. Quoted in Carter, p. 114.
6. Quoted in McDonald, p. 37.
7. Quoted in Carter, p. 116.
8. Quoted in Carter, p. 117.

→ V ←

Raising More Capital

1. Quoted in Judson, p. 68.
2. Quoted in Field, p. 81.
3. Quoted in Clarke, p. 35
4. Quoted in Clarke, p. 39.
5. Quoted in Carter, p. 126.

6. Quoted in Clarke, p. 37.

7. Quoted in Field, p. 103.

8. Field, p. 96.

9. Quoted in Field, p. 109.

⇥ VI ⇤
The First Cable

1. All quoted in Stampp, pp. 211–12.

2. Quoted in Bright, p. 47.

3. Quoted in Clarke, p. 40.

4. Bright, p. 47.

5. Quoted in Dibner, p. 20.

6. Quoted in Field, pp. 127–28.

7. Quoted in Field, pp. 129–30.

8. Quoted in Field, p. 131.

9. Field, p. 133.

10. Quoted in Field, p. 136.

11. Quoted in Bright, p. 70.

12. Quoted in Field, p. 138.

13. Quoted in Carter, p. 140.

14. Quoted in Bright, p. 71.

15. Quoted in Field, p. 137.

16. Quoted in Field, p. 139.

⇥ VII ⇤
"And Lay the Atlantic Cable in a Heap"

1. Quoted in Field, p. 144.

2. Field, p. 147.

3. Quoted in Dibner, p. 27.

4. Quoted in Judson, p. 87.

5. Field, p. 155.

6. Bright, p. 91.

7. Bright, p. 91.

8. Quoted, as are the following quotes on the storm, in Bright, pp. 91–105.

9. Quoted in Carter, p. 152.

10. Quoted in Field, p. 159.

11. Quoted in Field, p. 159.

12. Quoted in Carter, p. 153.

13. Quoted in Field, p. 161.

14. Quoted in Bright, p. 111.

VIII
Lightning Through Deep Waters

1. Field, p. 163.

2. Quoted in Judson, p. 92.

3. Quoted in Carter, p. 156.

4. Quoted in Bright, p. 117.

5. *North American Review,* April, 1866.

6. Field, p. 166.

7. Quoted in Bright, p. 118.

8. Quoted in Field, p. 169.

9. Quoted in Judson, p. 92.

10. Quoted in Bright, p. 121.

11. Quoted in Bright, p. 121.

12. Quoted in Bright, p. 124.

13. Quoted in Clarke, p. 57.

14. Quoted in Bright, p. 126.

15. Quoted in Bright, p. 127.

16. Quoted in Bright, p. 127.

17. Quoted in Bright, p. 130.

18. Quoted in Carter, p. 159.

19. Quoted in Bright, p. 132.

20. Quoted in Field, p. 172.

21. Quoted in Field, p. 172.

22. Quoted in Judson, p. 94.

23. Quoted in Judson, p. 95.

24. Quoted in Carter, p. 160.

25. Quoted in Field, p. 174.

26. Nevins, vol. II, pp. 407–8.

27. Nevins, vol. II, pp. 408–9.

28. Quoted in Carter, p. 161.

29. Quoted in Clarke, p. 60, and Carter, p. 165.

30. Quoted in Carter, p. 161.

31. Quoted in Carter, p. 161.

32. Quoted in Judson, pp. 100–101.

33. Quoted in Field, p. 197.

34. Nevins, vol. II, p. 410.

35. Quoted in Clarke, p. 63.

36. Quoted in Carter, p. 177.

37. Quoted in Carter, p. 178.

⇒ IX ⇐
"Here's the Ship to Lay Your Cable, Mr. Field"

1. Quoted in Clarke, p. 71.

2. Quoted in Clarke, pp. 70–71.

3. Quoted in Clarke, p. 70.

4. Quoted in Clarke, pp. 74–75.

5. Quoted in Clarke, p. 74.

6. Quoted in Rolt, p. viii f.

7. Quoted in Clark, p. 124.

8. Quoted in Clark, p. 180.

9. Quoted in Dugan, p. 6.

10. Quoted in Dugan, p. 1.

11. Quoted in Dugan, p. 73.

⇒ X ⇐
A New Cable, a New Attempt

1. Quoted in Foote, vol. I, p. 161.

2. Quoted in Carter, p. 192.

3. Quoted in Carter, p. 192f.

4. Quoted in Judson, p. 137.

5. Quoted in Carter, p. 196.

6. Field, p. 238.

7. Field, p. 239.

8. Field, p. 239.

9. Field, p. 240.

10. Quoted in Judson, p. 172.

11. Quoted in Judson, p. 172.

12. Quoted in Carter, p. 213.

13. Quoted in Carter, p. 213.

14. Quoted in Carter, p. 216.

15. Quoted in Judson, p. 179.

16. Boatner, p. 11.

17. Quoted in Carter, p. 217.

18. Quoted in Carter, p. 224.

19. Quoted in Carter, p. 224.

20. Russell, p. 51.

21. Quoted in Carter, p. 228.

22. Quoted in Dugan, p. 172.

23. Quoted in Carter, p. 229.

24. Russell, p. 60.

25. Field, p. 278.

26. Quoted in Judson, p. 192.

27. Russell, pp. 77–8.

28. Quoted in Field, p. 284.

29. Field, p. 286.

30. Quoted in Field, p. 287.

XI

"The *Great Eastern* Looms All Glorious in the Morning Sky"

1. Quoted in Dugan, p. 185.

2. Quoted in Field, p. 294.

3. Field, p. 296.

4. Quoted in Judson, p. 197.

5. Quoted in Field, p. 304.

6. Quoted in Carter, p. 244.

7. Quoted in Carter, p. 243.

8. Quoted in Carter, p. 245.

9. Quoted in Carter, p. 246.

10. Field, p. 319.

11. Quoted in Field, p. 323.

12. Quoted in Field, p. 323.

13. Quoted in Field, p. 325.

14. Quoted in Carter, p. 248.

15. Quoted in Judson, p. 205.

16. Quoted in Judson, p. 206.

17. Quoted in Field, p. 333.

18. Quoted in Field, p. 338.

19. Field, p. 341.

20. Quoted in Field, p. 339.

21. Quoted in Carter, p. 250.

22. Field, p. 342.

23. Quoted in Field, p. 344.

24. Quoted in Carter, p. 252.

25. *Harper's Weekly*, August 16, 1866.

26. Quoted in Judson, p. 207.

27. Quoted in Carter, p. 253.

28. Quoted in Carter, p. 254.

29. Quoted in Clarke, p. 96.

30. Quoted in Carter, p. 256.

31. Quoted in Carter, p. 256.

32. Field, p. 371.

33. Quoted in Bright, p. 19.

BIBLIOGRAPHY

Adams, John. *Ocean Steamers: A History of Ocean-going Passenger Steamships.* London: New Cavendish Books, 1993.

Boatner, Mark Mayo, III. *The Civil War Dictionary.* New York: David McKay, 1959.

Bright, Charles. *The Story of the Atlantic Cable.* New York: D. Appleton and Company, 1903.

Burke, James. *Connections.* Boston, Massachusetts: Little, Brown and Company, 1978.

Burrows, Edwin G., and Mike Wallace. *Gotham: A History of New York City to 1898.* New York: Oxford University Press, 1999.

Carter III, Samuel. *Cyrus Field: Man of Two Worlds.* New York: G. P. Putnam's Sons, 1968.

Clark, Ronald W. *Works of Man: A History of Invention and Engineering from the Pyramids to the Space Shuttle.* New York: Viking, 1985.

Clarke, Arthur. *Voice Across the Sea.* New York: Harper & Brothers, 1958.

Dibner, Bern. *The Atlantic Cable.* Norwalk, Connecticut: Burndy Library, 1959.

Dugan, James. *The Great Iron Ship.* New York: Harper & Bros., 1953.

Field, Henry M. *The Story of the Atlantic Cable.* New York: Charles Scribner's Sons, 1892.

Gordon, John Steele. *The Scarlet Woman of Wall Street.* New York: Weidenfeld and Nicolson, 1988.

Houghton, Walter E. *The Victorian Frame of Mind.* New Haven, Connecticut: Yale University Press, 1957.

Hunt, Bruce J. "Insulation for an Empire: Gutta-Percha and the Development of Electrical Measurement in Victorian Britain," in *Semaphores to Short Waves: Proceedings of a Conference on the Technology and Impact of Early*

Telecommunications, edited by Frank A. J. L. James. London: Royal Society for the Encouragement of Arts, Manufactures and Commerce, 1996.

Judson, Isabella Field, ed. *Cyrus W. Field, His Life and Work.* New York: Harper & Brothers, 1896.

McDonald, Philip B. *A Saga of the Seas: The Story of Cyrus Field and the Laying of the Atlantic Cable.* New York: Wilson-Erickson, 1937.

Meynell, Laurence. *Builder and Dreamer: A Life of Isambard Kingdom Brunel.* London: The Bodley Head, 1952.

Marvin, Carolyn. *When Old Technologies Were New: Thinking About Electric Communication in the Late Nineteenth Century.* New York: Oxford University Press, 1988.

Mullaly, John. *A Trip to Newfoundland.* New York: T. W. Strong, 1855.

———. *The Laying of the Cable, or the Ocean Telegraph.* New York: D. Appleton and Co., 1858.

Nevins, Allan, and Milton Halley Thomas (eds.). *The Diary of George Templeton Strong.* New York: The Macmillan Company, 1952.

Rolt, L. T. C. *Isambard Kingdom Brunel: A Biography.* London: Book Club Associates, 1959.

Royle, Trevor. *Crimea: The Great Crimean War 1854–1856.* New York: St. Martin's Press, 2000.

Russell, William Howard. *The Atlantic Telegraph.* London: Day & Son, 1866.

Shiers, George (ed.). *The Electric Telegraph: An Historical Anthology.* New York: Arno Press, 1977.

Stampp, Kenneth M. *America in 1857: A Nation on the Brink.* New York: Oxford University Press, 1990.

Standage, Tom. *The Victorian Internet: The Remarkable Story of the Telegraph and the Nineteenth Century's On-line Pioneers.* New York: Walker and Company, 1998.

Thompson, Robert Luther. *Wiring a Continent: The History of the Telegraph Industry in the United States.* Princeton, New Jersey: Princeton University Press, 1947.

Voorsanger, Catherine Hoover, and John K. Howat. *Art and the*

Empire City: New York, 1825–1861. New York: The Metropolitan Museum of Art, 2000.

Weintraub, Stanley. *Uncrowned King: The Life of Prince Albert.* New York: The Free Press, 1997.

Williams, Frances Leigh. *Matthew Fontaine Maury: Scientist of the Sea.* New Brunswick, New Jersey: Rutgers University Press, 1963.

INDEX